Reconfigurable RF and Microwave Technologies: Materials, Techniques, and Integration

Published 2025 by River Publishers
River Publishers
Alsbjergvej 10, 9260 Gistrup, Denmark
www.riverpublishers.com

Distributed exclusively by Routledge
605 Third Avenue, New York, NY 10017, USA
4 Park Square, Milton Park, Abingdon, Oxon OX14 4RN

Reconfigurable RF and Microwave Technologies: Materials, Techniques, and Integration / by Jinqun Ge, Guoan Wang.

Routledge is an imprint of the Taylor & Francis Group, an informa business

ISBN 978-87-7004-788-3 (paperback)

ISBN 978-87-7004-790-6 (online)

ISBN 978-87-7004-789-0 (ebook master)

A Publication in the River Publishers Series in Rapids

Reconfigurable RF and Microwave Technologies: Materials, Techniques, and Integration

Jinqun Ge

Qorvo US, Inc., USA

Guoan Wang

University of South Carolina, USA

River Publishers

R Routledge
Taylor & Francis Group

NEW YORK AND LONDON

Contents

Preface

The evolution of wireless communication systems has sparked an ever-increasing demand for more efficient, adaptable, and miniaturized RF and microwave components. As these systems become more complex, the need for reconfigurable technologies that can adapt to changing conditions and requirements has become paramount. This book, "Reconfigurable RF and Microwave Technologies: Materials, Techniques, and Integration," aims to explore these cutting-edge advancements, providing a comprehensive resource for researchers, engineers, and students in the field.

Chapter 1: Introduction, sets the stage by discussing the importance of reconfigurable RF components, the various manufacturing techniques employed, and the diverse tuning solutions available. This chapter also provides an overview of the book's organization, guiding readers through the subsequent sections.

Chapter 2: Fundamental of RF and Microwave Components, delves into the essential elements of RF and microwave systems, focusing on filters, antennas, and phase shifters. These components form the backbone of modern communication systems, and understanding their fundamentals is crucial for any advancement in reconfigurable technologies.

Chapter 3: Novel Fabrication and Manufacturing Techniques, highlights the innovative methods that have revolutionized the production of RF and microwave components. From micromachining techniques and 3D printing technologies to photolithography and thin film growing methods, this chapter explores the cutting-edge manufacturing processes that enable the creation of sophisticated and reliable components.

Chapter 4: Reconfigurable RF and Microwave Technologies, investigates the various approaches to achieving reconfigurability in RF and microwave systems. This includes mechanical tuning methods, semiconductor and RF MEMS switches, and advanced thin film technologies. These technologies offer the flexibility and adaptability required to meet the dynamic needs of modern communication systems.

Throughout this book, our goal is to provide a balanced mix of theoretical knowledge and practical insights, helping readers to not only understand the underlying principles but also apply them in real-world scenarios. We hope this book serves as a valuable reference for those seeking to enhance their understanding of reconfigurable RF and microwave technologies and drive innovation in this exciting field.

We extend our heartfelt gratitude to the many colleagues, researchers, and industry experts who have contributed their knowledge and experience to this work. Without their invaluable input, this book would not have been possible. We also thank our families for their unwavering support and encouragement throughout the writing process.

Thank you for embarking on this journey with us. We are confident that the insights and knowledge contained within these pages will inspire new ideas and advancements in the field of reconfigurable RF and microwave technologies.

Jinqun Ge and Guoan Wang

7 July 2024

About the Authors

Jinqun Ge, received his Ph.D. degree from the University of South Carolina, Columbia, SC, USA. His research interests include microwave circuits and components, surface micromachining technologies, ferromagnetic and ferroelectric materials, and smart electrically tunable microwave components. Dr. Ge was a recipient of the 2021 MTT-S Graduate Student Fellowship Award, Best Student Paper Award at the 2020 IEEE MTT-S International Wireless Symposium, and Student Paper Honorable Mention at the IEEE 2020 Antenna and Propagation Symposium.

He joined Qorvo US, Inc. in 2023 where he is working on BAW filter design. Dr. Ge is a reviewer for IEEE Microwave and Wireless Components Letters, International Journal of RF and Microwave Computer-Aided Engineering, Journal of Electronic Testing, International Journal of Antennas and Propagation, Physica Scripta, etc.

Guoan Wang received his Ph.D. in Electrical and Computer Engineering from Georgia Institute of Technology in 2006. He is currently a Professor in the Department of Electrical Engineering at the University of South Carolina. He worked as an Advisory Scientist responsible for the development of on chip mmwave passives and wafer level RF MEMS technologies at the IBM Semiconductor Research and Development Center from 2006–2011. His research areas include reconfigurable RF and microwave electronics, novel materials/techniques for smart RF applications, MEMS/NEMS, sensors and sensing systems, wireless energy harvesting, and 3D integrated devices/system. Dr. Wang's research work has produced over 140 papers in peer-reviewed journals and conferences proceedings. He also has 62 granted US and international patents, and 52 pending patent applications.

Dr. Wang served as an Associate Editor of IEEE Microwave and Wireless Components Letters from 2013–2019, and currently is an Associate Editor of

International Journal of RF and Microwave Computer-Aided Engineering. He is a member of the Technical Coordinating Committee for IEEE MTT TC-6 RF MEMS and Microwave Acoustics, chair of IEEE MTT TC-13 Microwave Control Materials and Devices, and vice-chair of IEEE MTT TC-17 Microwave Materials and Processing Technologies Committee. Dr. Wang is a recipient of the NSF Early Faculty Development (CAREER) Award in 2012, and IEEE Region 3 Outstanding Engineer Award in 2018.

1

Introduction

Over the years, wireless communication technologies have witnessed substantial advancements and have been extensively applied in various fields, including telecommunications, transportation systems, smart agriculture, wearable devices, military applications, satellite communication, and more [1, 2]. The fast-growing wireless communications market has and will continuously see dramatic changes in both the requirements and capabilities of radios to support wireless connections, especially in 5G and beyond technologies. Ultra-reliable and low-latency 5G wireless communication systems are required to accommodate over 50 frequency bands, each presenting its distinct characteristics, constraints, and specific challenges. Seamless roaming of mobile devices across geographic boundaries, adapting to diverse protocols and frequency bands as needed, is strongly desired. Tunable and reconfigurable RF technologies are enabling new frontiers for multiband and multiple function RF front ends, which significantly enhance radio spectrum accessibility by fostering interoperability among diverse standards, facilitating adaptation to environments with increased efficiency in spectrum utilization, power management, and device size [3, 4].

1.1 Reconfigurable RF Components

Traditional RF microsystems typically employ fixed-frequency transceivers optimized for performance within specific bands of interest. To accommodate multiple frequency bands and diverse wireless standards, separate RF front ends are commonly integrated in parallel, resulting in larger board areas and increased costs. A promising solution to achieve a simplified and miniaturized system involves the use of frequency-reconfigurable RF devices, as shown in Figure 1.1 [5]. In an ideal scenario, achieving communication within a specific frequency band becomes seamless by tuning the implemented reconfigurable RF components (e.g., filter and antenna) to the corresponding operating frequency range. Certainly, system complexity increases to maintain stable and high performance, especially when dealing with matching networks across various components and levels.

Figure 1.1: Schematic of a miniaturized multiple functional system enabled with frequency reconfigurable RF components.

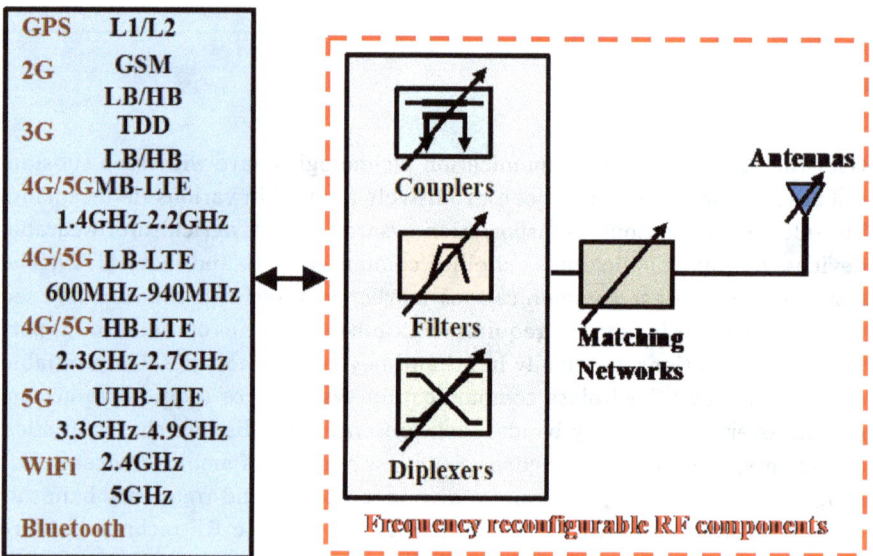

In the realm of reconfigurable RF components, a tunable filter stands out as a crucial element. Unlike its fixed-function counterparts, a tunable filter can dynamically adjust its frequency response. This adaptability proves valuable in scenarios where variable frequency ranges are encountered, offering a versatile solution for signal processing and communication systems. Another significant player in the world of tunable RF components is tunable antenna. Unlike conventional antennas, tunable antennas can dynamically adjust their resonance frequencies. This feature is particularly advantageous in environments with dynamic surrounding conditions where reconfiguration of the operating frequency is required to provide optimal performance. Tunable antennas contribute to enhanced signal reception and transmission, adapting to different communication standards and surrounding environments. To ensure optimal performance of tunable antennas and filters across diverse frequency bands, tunable matching networks are indispensable. These networks can be effectively implemented using tunable inductors and capacitors. Other reconfigurable devices, such as tunable phase shifters, tunable couplers, tunable amplifiers, etc., are also applicable to specific purposes. Overall, numerous studies propose various tunable RF devices today, and efficiently integrating these tunable devices into a communication system holds great promise for the development of intelligent wireless communication systems.

1.2 Manufacturing Techniques

Manufacturing techniques play a pivotal role in the development of high-performance RF components. The precision and quality achieved through various manufacturing processes directly impact the efficiency, reliability, and overall functionality of RF devices. Considerable endeavors have been dedicated to designing and implementing RF devices, utilizing a diverse array of manufacturing techniques:

- PCB (printed circuit board) manufacturing: Production of electronic circuit boards by etching copper-clad substrates, adding layers, and mounting electronic components. It is a fundamental step in the production of various electronic devices, from simple consumer gadgets to complex industrial equipment.
- Electroplating: A surface finishing technique that involves depositing a thin layer of metal onto a conductive substrate through an electrochemical process. When an electric current is applied, metal ions from the electrolyte are reduced and plated onto the substrate, creating a uniform and adherent metal coating. Electroplating is widely utilized for enhancing the corrosion resistance, conductivity, and aesthetics of objects.
- Photolithography [6]: A microfabrication technique that involves transferring a pattern onto a substrate using a photoresist and a photomask. This process involves applying a light-sensitive polymer, exposing it to ultraviolet light through a mask, and developing the exposed or unexposed areas, resulting in intricately defined structures essential for advanced electronic devices. It plays a critical role in creating intricate structures for integrated circuits and microelectromechanical systems (MEMS).
- Subtractive manufacturing: A process in which material is removed from a larger workpiece to achieve the desired shape or structure. Traditional machining techniques, such as milling, turning, and grinding, fall under subtractive manufacturing. It involves cutting, drilling, or shaping raw materials to create the final product, generating waste material in the process. Micromachining is a specialized form of subtractive manufacturing that focuses on producing extremely small-scale components with high precision. This technology facilitates the fabrication of highly integrated and compact RF devices.

- Additive manufacturing [7]: Commonly known as 3D printing, this is a manufacturing process that involves creating three-dimensional objects layer by layer. It utilizes digital models or computer-aided design (CAD) files to guide the additive manufacturing machine in depositing successive layers of material, which could be plastic, metal, or other substances, until the final object is formed. This approach contrasts with traditional subtractive manufacturing methods, such as machining or casting, as it adds material precisely where needed, reducing waste and enabling the production of complex and customized structures with greater efficiency and flexibility.
- Thin film growing [8]: A process for depositing thin layers of materials onto a substrate to create functional coatings or films. This technique typically involves physical vapor deposition (PVD) or chemical vapor deposition (CVD) methods, where atoms or molecules of the desired material are deposited onto a substrate surface under controlled conditions. Thin film growing is widely used in various industries, such as electronics and optics, to produce coatings with specific properties, including conductivity, reflectivity, or insulation, depending on the application.

Alternative manufacturing techniques, including extrusion, laser cutting, injection molding, and others, find applicability in specific contexts. From ensuring accurate fabrication of circuit boards using techniques like PCB manufacturing to the intricate detailing in components achieved through micro-machining, the application of appropriate manufacturing methods is crucial in meeting the demands of modern RF technologies. The careful selection and implementation of manufacturing techniques contribute significantly to optimizing the performance, miniaturization, and integration of RF components, ultimately shaping the advancement of wireless communication systems and related technologies.

1.3 Tuning Methods

The continually evolving manufacturing techniques have opened up greater possibilities for the design, implementation, and application of reconfigurable RF components. Moreover, novel materials are continually being developed to offer distinctive properties that can be harnessed for reconfigurable RF designs. Generally, the achievement of tunable RF devices involves the utilization of multiple technologies. The most common technology is RF switch technology, which includes

- Microelectromechanical systems (MEMS) [9]: Employing MEMS technology to create mechanically adjustable components in the RF structure.

- Semiconductor varactors: Utilizing the variable capacitance property of varactor diodes to achieve tunable capacitors and control the resonance frequency of the implemented components.
- Thin film (phase change material) switch: Implementing thin film switches, including phase change materials, to enable tunability.

In addition, novel thin film techniques, such as those utilizing ferroelectric and ferromagnetic materials, contribute to enhancing the performance of tunable RF devices by introducing unique electrical and magnetic properties [10]. Integrating ferroelectric or ferromagnetic thin films enables precise control over device characteristics, leading to improved efficiency and functionality. Besides RF switch technology and novel thin films, other complementary technologies also contribute to achieving RF tunability:

- Digital signal processing (DSP): Using digital techniques to dynamically adjust the characteristics of the RF component in real-time.
- Microstrip or stripline tunable components: Incorporating tunable components in microstrip or stripline configurations to achieve frequency agility.
- Liquid metal technology: Applying liquid metal elements to achieve tunability in certain types of RF devices.
- Graphene-based tunable devices: Exploring the unique properties of graphene to design tunable RF components with enhanced performance.

These technologies represent diverse approaches to achieve tunable RF components, each with its advantages and applications in specific scenarios. To address the demands of highly integrated and miniaturized modern applications, recent technological developments have focused on integrating tunable methods, providing continuous adjustability, ensuring broad applicability, and optimizing device performance. This book will delve into the technical aspects of reconfigurable RF technologies, covering material characteristics, integration techniques, the simultaneous application of multiple tuning methods, and presenting design examples.

1.4 Book Organization

In the journey through the realm of reconfigurable RF technology, this book unfolds a systematic exploration, delving into its foundational aspects, advanced processing techniques, and practical applications. Each chapter contributes uniquely to the comprehensive understanding of this evolving field.

Chapter 1 provides an overview of the background and context related to reconfigurable RF technology, setting the stage for understanding its significance and applications. The second chapter delves into foundational concepts, offering insights into essential RF components. It covers key principles of fundamental RF components. Chapter 3 explores cutting-edge processing techniques and fabrication technologies of RF components. This chapter summarizes the latest advancements in manufacturing processes. The fourth chapter is dedicated to an in-depth exploration of reconfigurable RF technologies, encompassing aspects such as design, integration, applications, and providing illustrative design examples. This chapter provides a practical understanding of the implementation and application of the developed components. The last chapter presents a comprehensive summary of the discussed topics and offers concise evaluations of the development and prospects within the field of reconfigurable RF technology. It concludes with insights into the future direction of this dynamic and evolving field.

References

[1]   Lyke J C, Christodoulou C G, Vera G A, et al. An introduction to reconfigurable systems. *Proceedings of the IEEE*, 103(3): 291-317, 2015.

[2]   Psychogiou D, Gómez-García R, Peroulis D. Recent advances in reconfigurable microwave filter design. *IEEE 17th Annual Wireless and Microwave Technology Conference (WAMICON)*. IEEE, 1-6, 2016.

[3]   Tan X, Zhang Y. Tunable Couplers: An Overview of Recently Developed Couplers with Tunable Functions. *IEEE Microwave Magazine*, 24(3): 20-33, 2023.

[4]   Zhang S, Zeng Q, Shang Y, et al. An overview of antenna reconfiguration technologies: Overview of Reconfigurable Antenna. *International Conference on Information Technology and Computer Application (ITCA)*, IEEE, 25-28, 2019.

[5]   Ge, J. *Novel Structures and Thin Film Techniques for Reconfigurable RF Technologies With Improved Signal Integrity*, 2023.

[6]   Lawson R A, Robinson A P G. Overview of materials and processes for lithography. *Frontiers of Nanoscience*, 11: 1-90, 2016.

[7]   Prakash K S, Nancharaih T, Rao V.S. Additive manufacturing techniques in manufacturing-an overview. *Materials Today: Proceedings*, 5(2): 3873-3882, 2018.

[8]   Zhang, Y., Ge, J., Wang, G. Enabling electrically tunable radio frequency components with advanced microfabrication and thin film techniques. *Journal of Central South University*, 29(10), 3248-3260, 2023.

[9]   Purtova T, Schumacher H. Overview of RF MEMS technology and applications. *Handbook of mems for wireless and mobile applications. Woodhead Publishing*, 3-29, 2013.

[10]   Ge, J., Wang, T., Peng, Y., & Wang, G. Electrically tunable microwave technologies with ferromagnetic thin film: Recent advances in design techniques and applications. *IEEE Microwave Magazine*, 23(11), 48-63, 2022.

2

Fundamentals of RF and Microwave Components

In the landscape of modern technology, the evolution of radio frequency (RF) and microwave technologies has been instrumental in shaping our interconnected world. From the early days of radio communication to the current era of advanced wireless networks, these technologies have become integral to our daily experiences, spanning telecommunications, broadcasting, radar systems, and a myriad of cutting-edge applications. Central to the chapter is a comprehensive examination of fundamental RF components, laying the groundwork for a deeper understanding of RF technology. Essential components such as filter, antenna, phase shifter, capacitor, and inductor are explored, each playing a crucial role in RF systems. By introducing the fundamental design principles of these components, readers will gain insights into their functionalities and contributions to RF systems. Real-world applications of these fundamental RF components are also discussed, showcasing their significance in diverse fields. Filters selectively pass or block frequencies, antennas facilitate signal transmission and reception, and phase shifters manipulate signal phase. These components are the backbone of wireless communication, satellite technology, and radar systems.

2.1 Filter

2.1.1 Filter basics

Filters are fundamental components in signal processing, electronics, and various engineering applications. They selectively pass or reject signals at certain frequencies, shaping the characteristics of signals, ensuring reliable transmission and reception of information in a spectrum of applications. There are four major types of filters, including low-pass filters, high-pass filters, band-pass filters, and notch filters (also known as a band-reject or band-stop filters). The terms "low" and "high" in this context do not denote absolute frequency values; instead, they are relative references to the cutoff frequency. Table 2.1 shows filtering responses of four types of filters, presenting the attenuation ratio (V_{out}/V_{in}) of the filter versus frequency. Attenuation is typically quantified in decibels (dB). Frequency can be represented in two formats: either the angular form ω (measured in rad/s) or the more conventional form of f (measured in Hz, i.e., cycles per second).

Table 2.1: Filtering responses of four main types of filters.

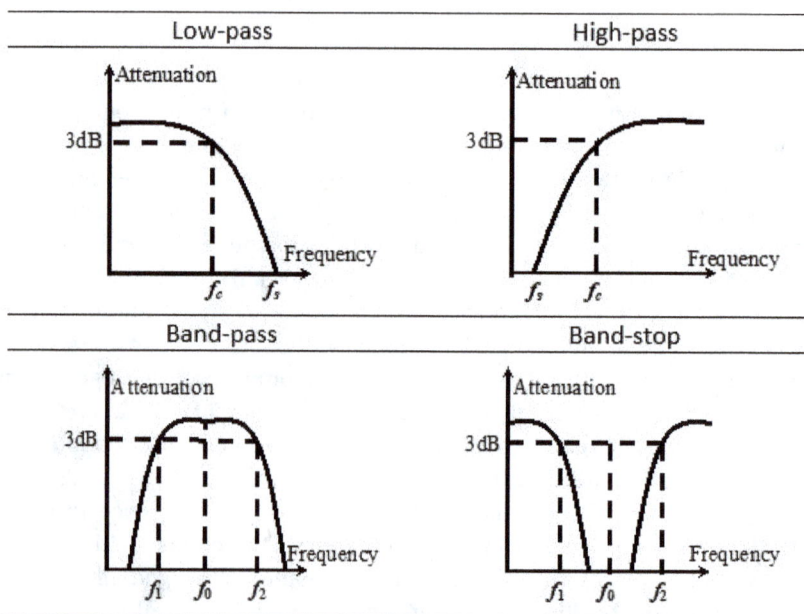

Filter performance is characterized by several key parameters:

- Center frequency ($f0$ or $\omega 0$): The frequency at which the filter exhibits its peak response or maximum attenuation.
- Bandwidth (BW): The range of frequencies around the center frequency where the filter's response remains significant. It is often defined as the difference between the upper and lower frequencies at which the response is reduced by a certain amount (e.g., -3 dB).
- Gain or attenuation: The magnitude of the filter's response, typically measured in dB. It indicates how much the filter amplifies or attenuates the input signal at different frequencies.

- Roll-off rate: Describes how quickly the filter's response transitions between the passband and stopband. It is usually expressed in decibels per octave or decibels per decade.
- Ripple: For some filters, particularly in the passband of an equalization filter, ripple represents variations in gain. Ripple is often expressed as a percentage of the filter's gain.
- Stopband attenuation: the level of attenuation in the stopband, indicating how effectively the filter suppresses unwanted frequencies. It is also specified in dB.
- Phase shift: the phase change of the filter's output signal compared to its input signal, often important in applications where phase relationships must be preserved.

These parameters collectively provide a comprehensive description of a filter's behavior and are crucial for selecting the right filter for a specific application.

2.1.2 Design and implementation

Filters can be designed and implemented through various methods. This section will focus on the most used lumped elements—capacitors and inductors—to illustrate how to design filters with different frequency responses. Additionally, it will introduce the most prevalent implementation methods and technologies for filters in use today.

Table 2.2 shows the types of filters formed by inductors and capacitors. Starting with the simplest component, the inductor, its impedance increases

Table 2.2: Four types of filtering responses with L and C.

Low-pass	High-pass

Band-pass	Band-stop

with frequency, effectively blocking high-frequency signals and forming a low-pass filter. Conversely, the impedance of a capacitor decreases with increasing frequency, behaving like an open circuit in direct current situations, allowing capacitors to form high-pass filters. When an inductor and capacitor are connected in series, a bandpass filter is created, simultaneously blocking both high and low-frequency signals. The operating frequency of this filter is the resonant frequency (f_0) where inductive reactance and capacitive reactance cancel each other out, expressed by the following formula:

$$f_0 = \frac{1}{2\pi\sqrt{LC}} \tag{2.1}$$

where L and C are the inductor and capacitor values. Similarly, if an LC circuit is connected in parallel, the impedance at the resonant frequency becomes infinite, forming a band-stop filter.

Combining these basic LC filtering structures allows the creation of filters with various filtering characteristics. Among them, band-pass filters are the most widely used. The ideal requirements for band-pass performance include low insertion loss, high out-of-band rejection, and sharp roll-off. Additionally, specific requirements for filter reliability, size, and other factors are tailored based on diverse application needs. Among the commonly used bandpass responses, the Butterworth bandpass response prioritizes a maximally flat response in the passband, sacrificing sharp roll-off for uniform amplitude. The Chebyshev bandpass response introduces passband ripple to achieve steeper roll-off in the stopband, offering a trade-off between flatness and roll-off. Additionally, the elliptic bandpass response combines features of Butterworth and Chebyshev filters, providing control over both passband ripple and stopband attenuation with a steeper roll-off. These response types offer a range of options for engineers to fine-tune filters based on considerations such as flatness, roll-off steepness, and passband ripple in accordance with specific design need.

Figure 2.1 (a) gives an example of a third-order elliptic filter example. It introduces three transmission poles and two transmission zeros, resulting in the filter response illustrated in Figure 2.1 (b). Compared to the Butterworth and Chebyshev filters, an elliptic filter can achieve a given set of specifications with a lower order, resulting in more compact design and reduced complexity. By increasing the filter order, additional poles and zeros can be introduced to improve in-band insertion loss, out-of-band rejection, and the roll-off response. However, high-order filters typically require more inductors and capacitors, leading to larger filter size and higher cost. While additional poles and zeros can improve certain aspects of filter performance, they may also introduce complexities, such as unwanted resonances, increased sensitivity to parasitic effects, and potential instability. Generally, increasing the order of an elliptic bandpass filter involves a trade-off between performance and complexity. A good filter design should consider the desired filter specifications, size constraints, and the available resources for implementation.

Figure 2.1: (a) Schematic of a third order elliptic bandpass filter and (b) its frequency response.

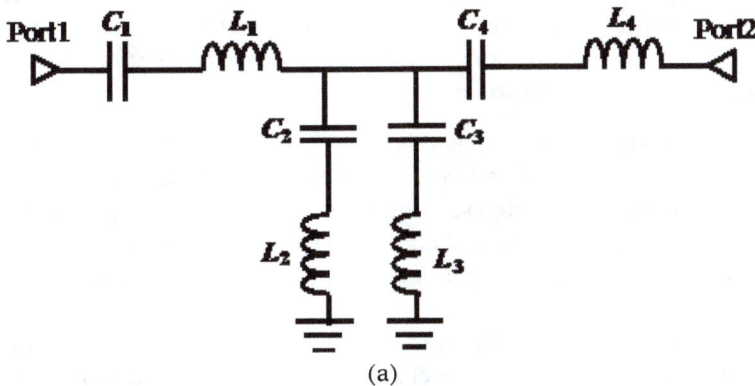

(a)

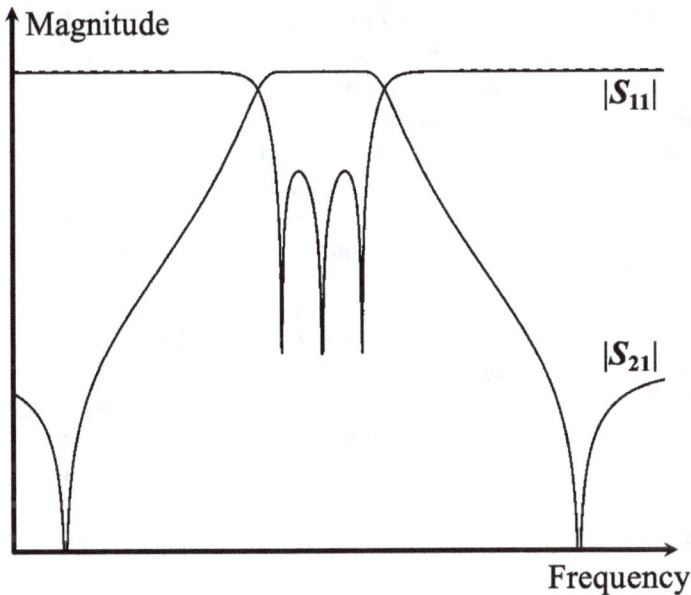

(b)

Traditional filters implemented with lumped elements have several draw-backs that may not meet the demands of modern applications. These short-comings include larger size and weight, lower Q-values, lower frequencies, and the presence of numerous parasitic effects. Despite these limitations, under-standing filters composed of lumped elements is crucial for designing other forms of filters. Various types of filters can often be theoretically analyzed by establishing equivalent circuits with lumped elements. This theoretical analysis aids in designing and optimizing filter performance. While traditional lumped element filters may have limitations, their fundamental principles serve as a valuable foundation for exploring and developing more advanced filtering solutions in modern applications.

As frequencies of mass-market designs push beyond the hundreds of MHz into the GHz range for applications such as smartphones, surface acoustic waves (SAWs) emerge as a promising solution for filter design. They operate by utiliz-ing acoustic waves propagating along the surface of a piezoelectric substrate to selectively filter and modify the frequency characteristics of signals. For appli-cations where precise frequency control is essential, temperature-compensated surface acoustic wave (TCSAW) filters are commonly used. Venturing into the realm of higher frequencies, especially in the context of 5G communication systems, bulk acoustic wave (BAW) filters have surged ahead, propelled by a host of advantages over their SAW counterparts. Their remarkable attributes, such as lower insertion loss, heightened temperature stability, and superior performance in compact designs, have catalyzed a swift and robust development in the landscape of filtering technology. This evolution underscores the dynamic shift toward BAW filters as key components, meeting the demands of advanced communication systems and ushering in a new era of efficient and reliable signal processing. The typical applicable frequency ranges of these filters are illustrated in Figure 2.2.

Figure 2.3 (a) shows a simplified BAW resonator structure, consisting of a layer of piezoelectric material sandwiched between two metal electrodes. Unlike SAW filters, the acoustic wave in a BAW filter propagates vertically, bouncing between the top and bottom electrode surface to form a standing acoustic wave. The resonance frequency is determined by the thickness of the piezoelectric material layer, as well as the thickness and mass of the electrodes. The electrical behavior of a BAW resonator can be approximately captured by an equivalent circuit using the Butterworth–VanDyke (BVD) model as shown in Figure 2.3 (b). The equivalent complex impedance is given by

$$Z_e = R_s + \frac{\left(j\omega L_m + \frac{1}{j\omega C_m} + R_m\right)\left(\frac{1}{j\omega C_0} + R_0\right)}{j\omega L_m + \frac{1}{j\omega C_m} + \frac{1}{j\omega C_0} + R_m + R_0} \tag{2.2}$$

Figure 2.2: Application frequency ranges of three types of acoustic filters.

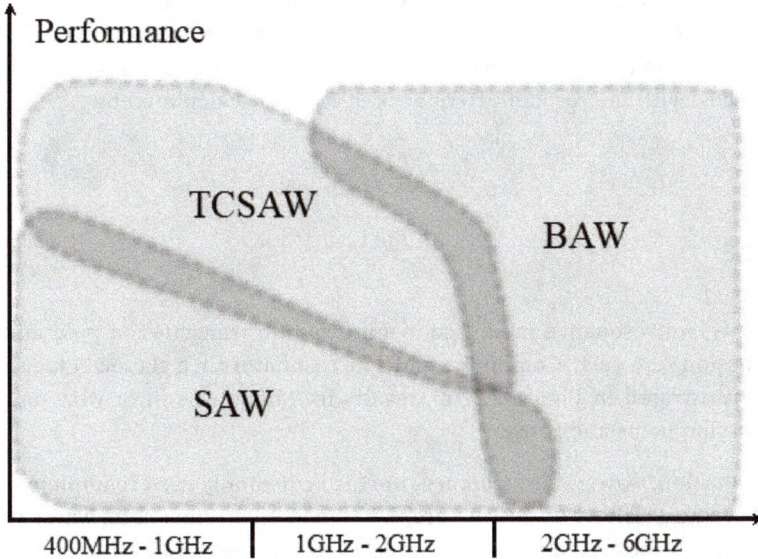

Performance

TCSAW

BAW

SAW

400MHz - 1GHz | 1GHz - 2GHz | 2GHz - 6GHz

Figure 2.3: (a) Schematic of a bulk acoustic wave (BAW) resonator. (b) The equivalent circuit of a BAW resonator. (c) Topology of a BAW ladder bandpass filter.

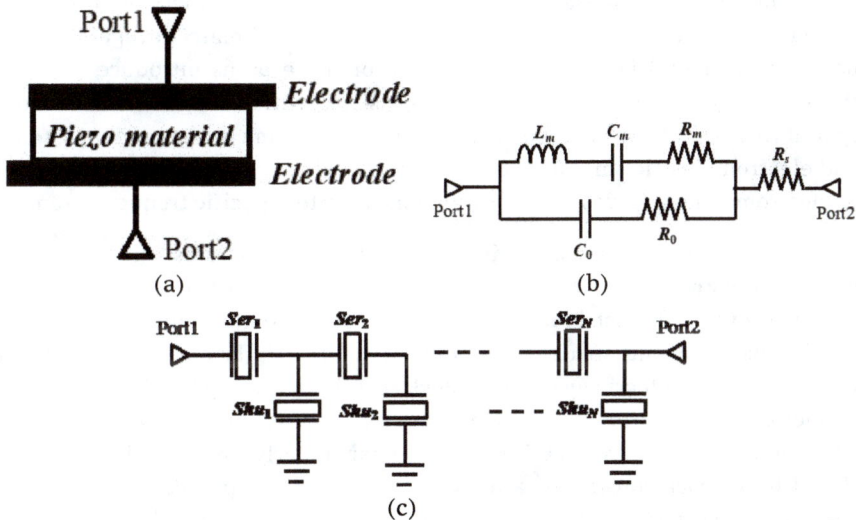

Port1

Electrode

Piezo material

Electrode

Port2

(a)

L_m C_m R_m

R_s

Port1

C_0 R_0

Port2

(b)

Port1 Ser_1 Ser_2 Ser_N Port2

Shu_1 Shu_2 Shu_N

(c)

where L_m and C_m are the elements of the acoustic resonance frequency, R_m is the mechanical resistor of the piezoelectric layer, R_s is the resistor of the electrodes, C_0 is the physical and static capacitor, R_0 is the dielectric resistor of piezo material, and ω is the angular frequency of the harmonic signal flowing through the circuit. The BAW resonator exhibits both series and parallel resonances, with the resonant frequency of f_s and f_p calculated by

$$f_s = \frac{1}{2\pi\sqrt{L_m C_m}} \tag{2.3}$$

$$f_p = \frac{1}{2\pi}\sqrt{\frac{C_m + C_0}{C_m C_0 L_m}}. \tag{2.4}$$

These two resonance modes contribute to one transmission pole and one transmission zero each. Combining multiple resonators in a classic ladder topology, as illustrated in Figure 2.3 (c), results in a bandpass filter with multiple transmission poles and zeros.

In addition, waveguide structures are also commonly used to implement filters. Waveguide filters utilize the propagation of electromagnetic waves within waveguide structures to create filters for specific frequency bands, and their design and construction is dependent on the desired frequency characteristics. Due to the inherent properties of waveguides, these filters exhibit high Q, low insertion loss, and high power handling capabilities. Figure 2.4 illustrates a spatial filter implemented with waveguides, known as a frequency selective surface (FSS) [2]. It is composed of periodic structures, with each unit cell consisting of three square coaxial tubes. The structure provides two square coaxial waveguide propagation paths and one parallel-plate waveguide patch between the metal layers. The resonances of these paths introduce multiple transmission poles and zeros, forming a passband with a shaped roll-off. Those spatial filters find applications in areas such as antenna design, radar systems, and electromagnetic shielding, allowing engineers to manipulate the behavior of electromagnetic waves in a controlled manner for specific frequency ranges.

Another crucial approach for implementing filters is microstrip line technology. A microstrip line is a type of transmission line consisting of a conducting strip laid over a dielectric layer and backed by a ground plane. By designing specific patterns and structures within the microstrip layout, these lines can form filters that selectively allow or block certain frequency bands. Advantages of microstrip line filters include compact size, low cost, and ease of integration on printed circuit boards (PCBs). They are particularly well-suited for applications where space constraints are critical, making them popular in various RF and microwave systems. Figure 2.5 gives an example of a dual-band bandpass filtering balun achieved with microstrip lines. This filter design comprises two

Figure 2.4: Frequency selective surface by waveguides [2].

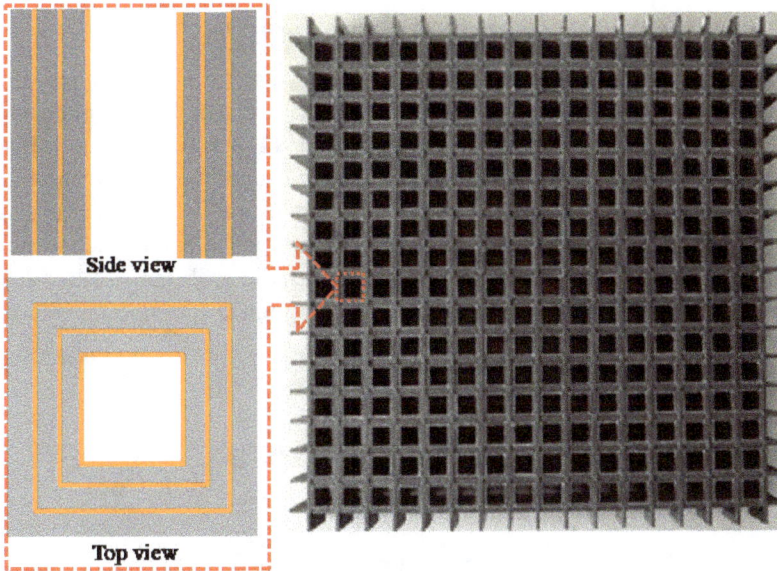

Figure 2.5: Structure and optical photo of a bandpass microstrip line filter [3].

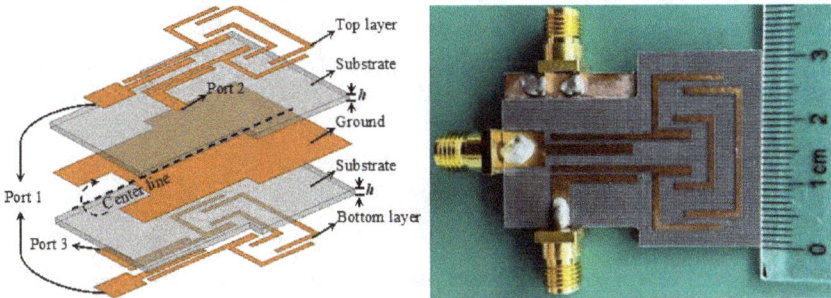

symmetrical filtering structures with double-sided parallel-strip lines incorporated into the input port. The filtering structure itself is composed of two half-wavelength resonators and two open-stub loaded resonators, yielding two third-order passbands. A specially developed coupling scheme introduces multiple transmission zeros, achieving an exceptionally sharp roll-off, ideal for highly selective filters. Furthermore, the double-sided parallel-strip lines contribute an out-of-phase feature, enabling the balun function. Generally, microstrip lines are employed to create specific resonators, such as half-/quarter-wavelength resonators, open-/short-stub loaded resonators, multiple-stub resonators, etc. These resonators introduce specific resonant modes to generate the desired passband filtering responses.

In addition to the common filter implementation methods mentioned above, there are also other forms of filters, such as crystal filters, ceramic filters, dielectric resonator filters, etc. The diverse landscape of filter technologies encompasses various approaches, each tailored to specific applications and performance requirements. From traditional lumped elements to advanced acoustic and waveguide/microstrip line structures, the evolution of filter design continues to meet the demands of modern communication systems and electronic devices.

2.2 Antenna

2.2.1 Antenna basics

Antennas serve as fundamental components in wireless communication systems, acting as interfaces between electromagnetic waves and electronic devices. These devices play a pivotal role in transmitting and receiving electromagnetic signals, enabling the wireless transmission of information across vast distances. Antennas come in various shapes and sizes, each designed to fulfill specific communication requirements in broadcasting radio and television signals, establishing cellular connections, and enabling satellite communication. As technology continues to advance, antennas remain essential in facilitating the ever-growing demand for wireless connectivity in our increasingly interconnected world.

Antennas radiate electromagnetic waves when an alternating current flows through them. This current causes the electrons in the antenna to accelerate, producing changing electric and magnetic fields. As these fields interact, they propagate away from the antenna as electromagnetic radiation, carrying energy and information through space. Understanding the basics of antennas is fundamental to understand their significance in transmitting and receiving electromagnetic signals effectively. Here are some essential concepts related to antennas:

- Radiation pattern: Describes the directional properties of an antenna's radiation in three dimensions.

- Directivity (D): Describes the concentration of radiation in a particular direction relative to an isotropic radiator.
- Gain (G): Measures the directionality of an antenna's radiation pattern, indicating its ability to focus energy in a particular direction. Antenna gain (G) can be related to directivity (D) and antenna efficiency (ϵ_R) by $G = \epsilon_R D$.
- Polarization: Refers to the orientation of the electromagnetic fields produced by an antenna, which can be linear, circular, or elliptical.
- Resonant frequency: The frequency at which an antenna operates most efficiently, determined by its physical dimensions and electrical properties.
- Bandwidth: The range of frequencies over which an antenna can operate effectively without significant loss of performance.
- Radiation efficiency: Indicates the percentage of input power that an antenna converts into radiated energy.
- Impedance: Represents the opposition to the flow of alternating current in an antenna, crucial for efficient power transfer between the antenna and the transmission line. The fundamentals of antenna theory require that the antenna be "impedance matched" to the transmission line or the antenna will not radiate.
- Voltage standing wave ratio (VSWR): Measures the efficiency of power transfer between the transmission line and the antenna, with lower values indicating better matching.

2.2.2 Design and implementation

Starting with one of the simplest antennas—the dipole antenna—which is widely used due to its balanced radiation pattern and relatively straightforward construction. As shown in Figure 2.6 (a), a dipole antenna consists of two conductive elements, typically metal rods or wires, oriented in parallel and separated by a small gap. It is simply an open-circuited wire, fed at its center. Dipole antennas find common applications in radio and television broadcasting, as well as various wireless communication systems. They operate efficiently over a broad range of frequencies and are relatively easy to install and tune.

The fundamental resonance of a thin linear conductor occurs at a frequency where the free-space wavelength is twice the length of the wire, meaning the conductor is half of the wavelength long. Dipole antennas are commonly utilized at this frequency and are therefore termed half-wave dipole antennas. Figure 2.6 (b) shows the radiation patterns of a half-wavelength dipole antenna. The far-fields from a dipole antenna of length L are given by:

$$E_\theta = \frac{j\eta I_0 e^{-jkr}}{2\pi r} \left[\frac{\cos\left(\frac{kL}{2}\cos\theta\right) - \cos\left(\frac{kL}{2}\right)}{\sin\theta} \right] \qquad (2.5)$$

$$H_\emptyset = \frac{E_\theta}{\eta} \qquad (2.6)$$

Figure 2.6: (a) A center-fed half-wavelength dipole antenna and its radiation patterns. (b) The radiation patterns of a half-wavelength dipole antenna.

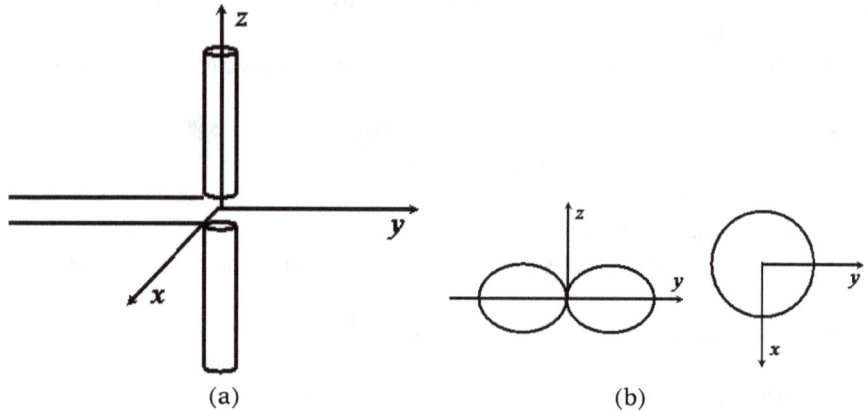

(a)   (b)

where η is the intrinsic impedance of free space, k is the wavenumber, I_0 is the feeding current amplitude, θ denotes the direction, and r is the distance between calculated point and the antenna center point.

After discussing the characteristics of a dipole antenna, it's important to consider its variant, the monopole antenna. Unlike dipole antennas, monopole antennas utilize a single conductive element placed above a conducting ground plane, as illustrated in Figure 2.7 (a). This configuration makes it particularly suitable for applications where space is limited or where a ground plane is readily available, such as handheld devices and vehicles. Using image theory, the fields above the ground plane mirror those of a dipole antenna with twice the length, while the fields below the ground plane are zero, as shown in Figure 2.7 (b). This principle allows for the design and analysis of monopole antennas by leveraging the known behavior of dipole antennas. The key difference lies in the impedance: a monopole antenna has half the impedance of a full dipole antenna. This discrepancy arises because a monopole requires only half the voltage to produce the same current as a dipole. In a dipole, there are $+V/2$ and $-V/2$ voltages applied to its ends, whereas a monopole antenna needs only $+V/2$ voltage between itself and the ground to achieve an equivalent current. Therefore, the impedance of the monopole antenna is halved compared to that of the dipole.

The performance of a dipole antenna can be enhanced by adding a reflector and several directors, forming a Yagi antenna, as shown in Figure 2.8 (a). The

Figure 2.7: (a) A monopole antenna and (b) its radiation patterns.

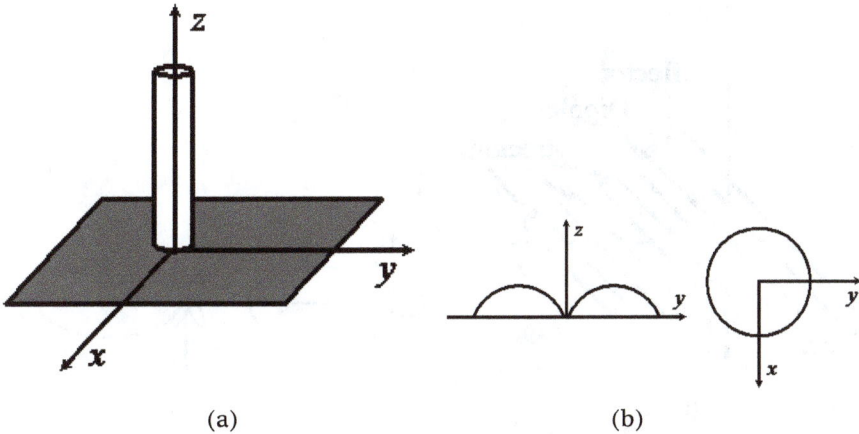

(a)   (b)

dipole serves as the driven element, with its length typically ranging from 0.45λ to 0.48λ to ensure resonance in the presence of the parasitic elements. The reflector is positioned behind the drive element and typically has a length of around 5% longer than the driven element. Its spacing from the drive element at about 0.1λ to 0.5λ, with the specific spacing dependent on factors such as bandwidth, gain, front-to-back ratio, and sidelobe pattern requirements. Directors are primarily intended for receiving electromagnetic waves. The closest director to the dipole is approximately 5% shorter than the driven element, and subsequent directors decrease in length sequentially. Figure 2.8 (b) shows the typical radiation pattern of a Yagi antenna. This design creates directional gain, improving the antenna's reception and transmission capabilities in specific directions. It has moderate to high gain of up to 20 dBi, depending on the number of elements used. The Yagi antenna's ability to focus its radiation pattern makes it particularly effective for long-distance communication and reception, especially in scenarios where signals need to be transmitted or received from a specific direction. Its relatively simple construction, combined with its high performance and cost-effectiveness, has led to widespread adoption in various communication systems worldwide. Additionally, Yagi antennas are commonly used in amateur radio operations, wireless internet access, and point-to-point communication links, showcasing their versatility across different applications and industries.

Another commonly used antenna is the patch (microstrip) antenna, which is known for its low profile, lightweight design, and ease of integration with

Figure 2.8: (a) A Yagi antenna and (b) its radiation patterns.

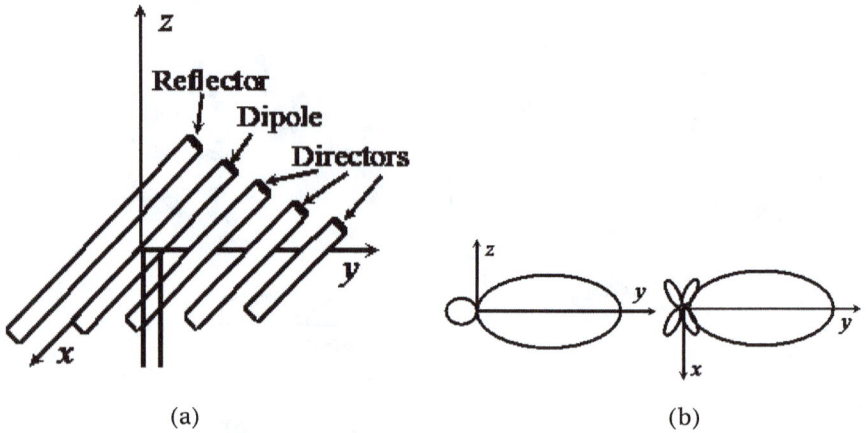

(a)

(b)

Figure 2.9: (a) A patch antenna and (b) its radiation patterns.

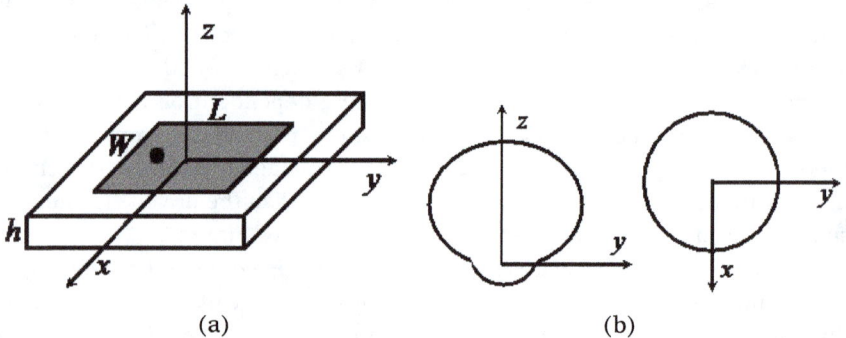

(a)

(b)

electronic devices. As depicted in Figure 2.9 (a), it consists of a radiating patch, a substrate, and a ground plane. The patch antennas can be fed by a microstrip line, a coaxial cable, aperture coupling, proximity coupling, or inset feeding. The length of the patch antenna (L) determines its operating frequency (f_0) while the width of the antenna (W) controls the input impedance. Additionally, wider widths can enhance the antenna's bandwidth.

The design equations of the patch antenna are given as

$$W = \frac{c}{2f_0\sqrt{\frac{(\varepsilon_r+1)}{2}}} \tag{2.7}$$

$$L = L_{eff} - 2\Delta L \tag{2.8}$$

$$L_{eff} = \frac{c}{2f_0\sqrt{\varepsilon_{eff}}} \tag{2.9}$$

$$\Delta L = 0.412h \frac{(\varepsilon_{eff} + 0.3)\left(\frac{W}{h} + 0.264\right)}{(\varepsilon_{eff} - 0.258)\left(\frac{W}{h} + 0.8\right)} \tag{2.10}$$

where ϵ_r is the dielectric constant of the substrate and c is the speed of light in free space. They are commonly used in wireless communication systems, such as WiFi routers, mobile phones, and satellite communication terminals, due to their compact size and directional radiation pattern.

In addition to the basic antennas discussed above, there are several other fundamental antenna types commonly used in various applications. These include inverted F antennas, horn antennas, helical antennas, and slot antennas. Each of these antenna types has unique characteristics and advantages suitable for specific requirements. For instance, inverted F antennas offer compact designs ideal for integration into portable devices, while horn antennas provide high directivity and gain for microwave applications. Similarly, helical antennas are known for their circular polarization and omnidirectional radiation patterns, making them suitable for satellite communication and GPS systems. Slot antennas, on the other hand, are often employed in radar systems and wireless communication due to their versatility and ease of fabrication. Together, these antenna types form the foundational elements of modern wireless communication systems, catering to a diverse range of applications across various industries.

Figure 2.10: (a) A dual-function antenna with controllable EM and thermal performance. (b) 3D radiation pattern of the antenna.

(a)   (b)

Building upon these fundamental antenna designs, further customization and optimization are often performed to meet specific application requirements, enabling additional functionalities or improved performance. Figure 2.10 (a) shows a design example of a dual-function antenna that exhibits both good electromagnetic (EM) performance and thermal performance [4]. The non-contact heatsink is lifted with a tiny air gap from the antenna and partially connected with the antenna through physical switches/poles that balance the EM performance and heat dissipation efficiency, which eliminates the negative impact of the heatsink on the antenna EM performance. The overarching concept of this dual-function antenna involves connecting the heatsink to the selective area of the bottom antenna, where the E-field strength is weakest.

Understanding the different types of antennas and their characteristics is paramount to achieve the desired communication functionalities in antenna designs and applications. They provide the foundation for various communication functionalities and serve critical roles in different application domains. Moreover, these basic antennas lay the groundwork for further optimization and customization to meet the specific requirements of diverse application scenarios.

2.3 Phase Shifter

2.3.1 Basics of a phase shifter

A phase shifter is a crucial component in RF and microwave systems that allows for the precise control and manipulation of the phase of an electromagnetic signal. As illustrated in Figure 2.11, an RF phase shifter will shift the phase of the signal without varying the amplitude of the input signal, after considering the insertion loss of the component. It is commonly used in applications such as phased array antennas, radar systems, communication systems, and beamforming networks.

Figure 2.11: Phase difference between the input and output signal of a phase shifter.

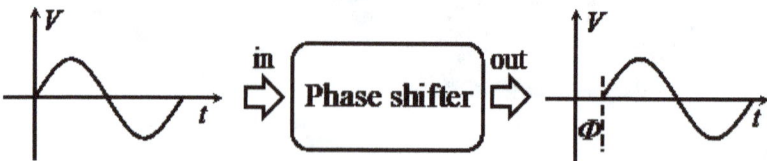

The performance of a phase shifter is typically evaluated based on several key parameters, including:

- Phase shift range: This refers to the range over which the phase shifter can vary the phase of the input signal. It is typically specified in degrees or radians.
- Insertion loss: This is the amount of signal power lost when passing through the phase shifter. Lower insertion loss indicates better performance.
- Return loss: This parameter measures the amount of power reflected towards the source due to impedance mismatches. Higher return loss indicates better impedance matching and less signal loss.
- Frequency range: The frequency range over which the phase shifter operates effectively without significant degradation in performance.
- Accuracy and linearity: These parameters measure how closely the actual phase shift matches the desired phase shift over the operating range. Higher accuracy and linearity indicate better performance.
- Power handling capability: This parameter indicates the maximum power level that the phase shifter can handle without degradation in performance or damage.

By evaluating these parameters, engineers can assess the suitability of a phase shifter for a particular application and determine its overall performance characteristics.

2.3.2 Design and implementation

Phase shifters can be realized mechanically or electrically. Mechanical phase shifters offer simple and robust design, making them suitable for certain applications despite being less common in modern electronic systems. They provide a tangible means of phase adjustment but may sacrifice speed compared to electronic alternatives. Mechanical phase shifters are often favored in applications where ruggedness and high power handling capabilities are paramount. Digital phase shifters, on the other hand, provide precise and repeatable phase adjustments through digital control circuits. They are favored in digital communication and phased array systems for their ability to offer discrete phase shifts with high accuracy. By utilizing digital control interfaces and sophisticated algorithms, digital phase shifters offer flexibility and adaptability in controlling phase, making them indispensable in modern communication networks and radar systems. Finally, analog phase shifters provide continuous adjustment of signal phase through electronic components like capacitors, inductors, or transmission lines, offering smooth and continuous phase tuning, albeit with potentially limited resolution. This chapter primarily focuses on analog phase shifters due to their widespread use across RF, microwave, and communication systems. It delves into their operational principles, design intricacies, and practical implementations to offer a comprehensive understanding of this crucial component in electronic signal processing.

Figure 2.12 shows a simple delay line (switched line) phase shifter. It consists of a reference arm, a delay arm, and two switches. The phase shift can be easily calculated by determining the difference in the electrical lengths of the reference arm and the delay arm. The phase of any transmission line is equal to its length multiplied by the propagation constant. Different phase shifts can be achieved by varying the length of the transmission line or by using techniques such as cascading multiple stages. By adjusting the delay element, the phase shift introduced to the signal can be precisely controlled over a certain range. Depending on specific application requirements, the transmission line can be chosen from options such as coaxial cable, microstrip line, waveguide, or stripline.

Figure 2.12: An example of a delay line (switched line) phase shifter.

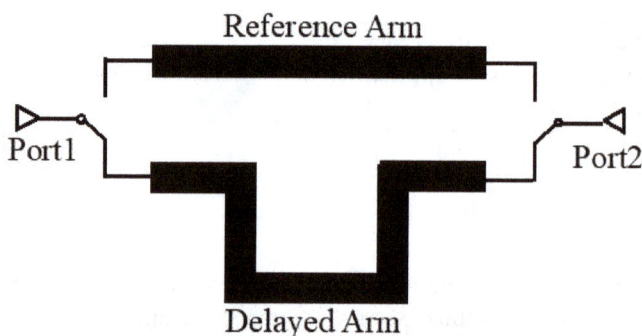

Delay line phase shifters are valuable components in electronic systems where precise control over a phase relationship is essential, offering versatility, reliability, and ease of integration in various applications. Nevertheless, delay line phase shifters can present certain drawbacks attributed to the characteristics of the transmission line employed. These drawbacks include limited bandwidth, sizable dimensions, and non-continuous phase control. Furthermore, the inherent losses and dispersion introduced by transmission lines, particularly notable at higher frequencies, can lead to compromised signal amplitude and distortion, ultimately affecting the overall performance of the system.

Another category of phase shifter is the loaded line phase shifter, which is often used for 45 degree or lower shift. As shown in Figure 2.13, it utilizes a transmission line loaded with reactive elements such as inductors or capacitors to introduce a phase shift in an electrical signal. Unlike delay line phase shifters, which rely on varying the physical length of the transmission line, loaded-line phase shifters manipulate the electrical properties of the line to achieve the desired phase shift. By strategically placing and tuning these reactive elements along the transmission line, precise and continuous phase adjustment can be achieved over a certain frequency range. The loads, denoted as Z_L, are specifically designed to induce a phase perturbation in the signal upon their activation within the circuit, while exerting minimal influence on signal amplitude. It is imperative for these loads to possess a considerably high reflection coefficient to mitigate phase shifter losses effectively. However, it's crucial to ensure that the loads Z_L are not positioned too closely to a short circuit in terms of phase angle, as this could lead to significant loss within the phase shifter mechanism. To optimize performance, spacing the reactive loads at approximately quarter-wavelength intervals serves to minimize and balance amplitude perturbations across both states. Typically, the phase versus frequency response of a loaded-line phase shifter exhibits greater flatness compared to that of a switched-line phase shifter. It also offers advantages such as compact size, wide bandwidth, and continuous phase adjustment, making them suitable for various RF and microwave applications where precise phase control is required. Additionally, loaded-line phase shifters can be designed with low insertion loss and high power handling capabilities, further enhancing their utilization in electronic systems.

Figure 2.13: An example of a loaded line phase shifter.

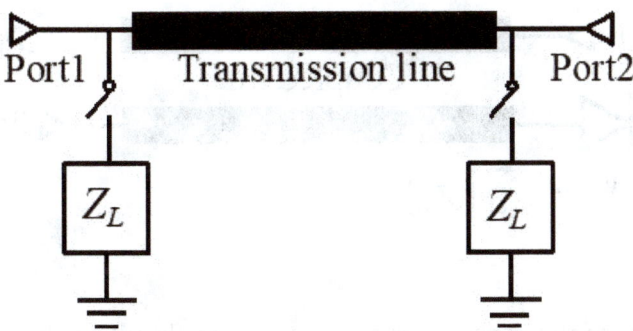

The third type of phase shifter is the reflective phase shifter, which operates on the principle of reflection, utilizing impedance mismatches to induce phase shifts. This type of phase shifter typically consists of transmission lines terminated with mismatched impedances, causing incident signals to reflect with altered phase characteristics. By adjusting the termination impedance utilizing varactors, the phase shift can be controlled over a certain frequency range. Reflective phase shifters offer advantages such as simplicity in design, wide bandwidth capability, and low insertion loss. Figure 2.14 gives an example of a reflective phase shifter, which consists of a section of coupled line and two varactors. Varactors are semiconductor devices whose capacitance is controllable with different applied voltage. By incorporating varactors into the phase shifter circuitry, the phase shift can be continuously adjusted over a certain range by varying the voltage applied to the varactors. This tunable approach allows for precise and flexible control of the phase shift, making it well-suited for applications where continuous phase adjustment is required, such as in phased array antennas and frequency-agile communication systems. The phase shift (Φ) of this reflective phase shifter is obtained from the phase variation of reflection coefficient (Γ) at the termination load (Z_{in}). The reflection coefficient can be calculated as:

$$\Gamma = |\Gamma| e^{j\Phi} = \frac{Z_{in} - Z_0}{Z_{in} + Z_0} \tag{2.11}$$

where Z_0 is the characteristic impedance of the coupled line. It's crucial to design the coupled line with tight coupling to achieve broadband performance, low insertion loss, and minimal phase error. Additionally, the length of the coupled line should be carefully optimized, considering the impact of varactor capacitance.

Figure 2.14: An example of reflective phase shifter enabled with varactors.

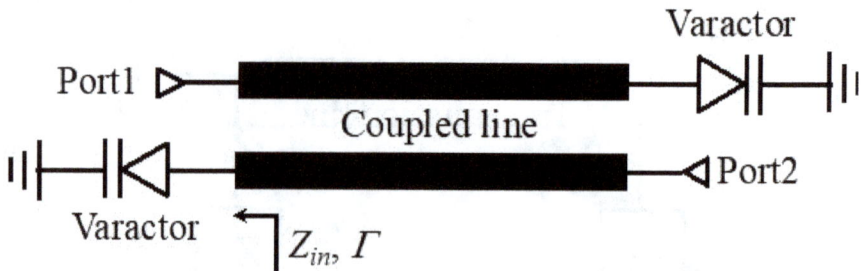

Beyond the mentioned types of phase shifters, there's another avenue utilizing unique properties of materials, exemplified by the ferroelectric phase shifter. A ferroelectric phase shifter utilizes ferroelectric materials, which exhibit a unique property called ferroelectricity. Figure 2.15 illustrates an example of a ferroelectric phase shifter comprising a microstrip line on a ferroelectric substrate. When subjected to an electric field generated by the bias voltage, ferroelectric materials can undergo reversible polarization changes,

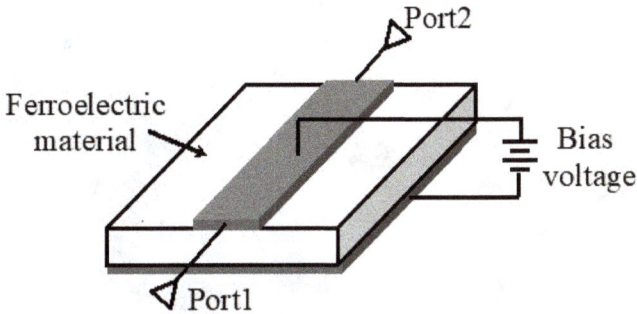

Figure 2.15: An example of a ferroelectric phase shifter.

resulting in variations in their dielectric constant. By modulating the electric field applied to the ferroelectric material, the phase shift of the transmitted signal can be controlled. This allows for fast and efficient phase modulation without the need for bulky mechanical components or complex circuitry. Additionally, ferroelectric phase shifters tend to consume relatively low power compared to some other types of phase shifters. They also have the potential for broad bandwidths, making them suitable for applications requiring frequency agility. However, ferroelectric phase shifters may exhibit a limited range over which they can effectively induce phase shifts, constraining their applicability in certain scenarios. Furthermore, the relationship between the applied electric field and the resulting phase shift in ferroelectric materials may be nonlinear, potentially introducing signal distortions.

There are also other varieties of phase shifters, including lumped element phase shifters, switched filter phase shifters, and circulator-based phase shifters. Each of these types offers unique advantages and applications in various RF and microwave systems. Understanding these basic types of phase shifters enables better design tailored to specific application requirements.

Nowadays, phase shifter design is not confined to a single technique or type, but rather incorporates multiple technologies to achieve more complex and superior performance phase shifters. A good example is depicted in Figure 2.16, which achieves a tunable phase shift response using ferroelectric and ferromagnetic materials. The foundational structure of this design is based on a slow wave coplanar waveguide structure with discontinuous signal width, as illustrated in Figure 2.16 (a). The narrow signal line and larger distance between the signal line and ground plane contributes to the high inductance section, whereas the wide signal line and small distance to ground contributes to the high capacitance section. When these two sections are cascaded together,

Figure 2.16: (a) Slow wave coplanar waveguide structure. (b) Tunable phase shifter enabled with ferroelectric and ferromagnetic materials.

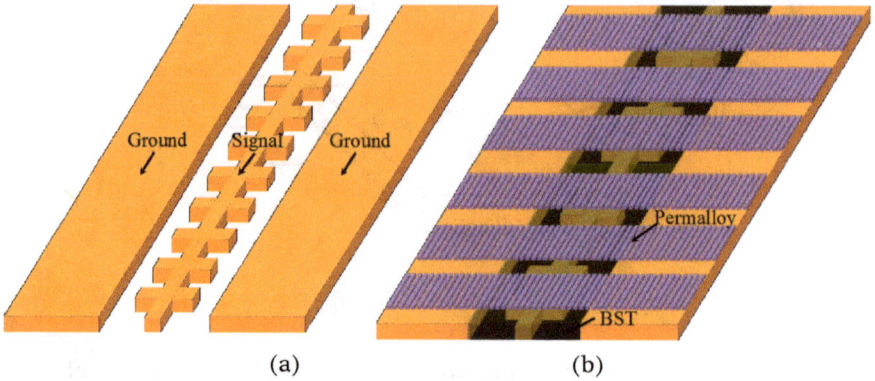

Ground Signal Ground

Permalloy

BST

(a)   (b)

the inductance per unit length of the line is primarily dominated by the high inductance section, while the capacitance per unit length of the line is dominated by the high capacitance section. As a result, there is a simultaneous increase in both the inductance and capacitance of the signal line, leading to an increased signal delay. Building upon this concept, as depicted in Figure 2.16 (b), the inductance and capacitance per unit length of the line can be augmented by incorporating ferromagnetic material (e.g., permalloy) and ferroelectric material (e.g., barium strontium titanate, BST), respectively. As the permeability and permittivity of these materials vary under different magnetic and electric fields, respectively, a tunable phase shift response can be achieved. The utilization of ferroelectric and ferromagnetic materials in tunable phase shifters facilitates dynamic and continuous adjustments in phase response. This capability enables precise control over phase shift characteristics, ultimately enhancing the versatility and performance of RF systems.

References

[1]   Pozar D M. *Microwave engineering*. John Wiley & sons, 2011.
[2]   W. Tang, J. Ge, Z. Yu, H. Lu and B. Li, 3-D waveguide FSS by coaxial square tubes. *2017 International Conference on Electromagnetics in Advanced Applications (ICEAA)*, Verona, Italy, 1746-1748, 2017.
[3]   J. Ge, and G. Wang, A dual-band filtering structure for highly selective reconfigurable bandpass filter and filtering balun. *International Journal of RF and Microwave Computer-Aided Engineering*, 32(8), e23215, 2022.

[4]   J. Ge, C. Li, and G. Wang, Investigation of a novel heat dissipation concept with controllable thermal and EM performance for reliable electronics and communication systems. *Engineering Research Express*, 5(2), 025014, 2023.

[5]   G. Wang, F. Rahman, T. Xia and H. Zhang, Patterned Permalloy and Barium Strontium Titanate Thin Film Enabled Tunable Slow Wave Elements for Compact Multi-Band RF Applications. *IEEE Transactions on Magnetics*, 49(7), 4184-4187, 2013.

CHAPTER

3

Novel Fabrication and Manufacturing Techniques

In the realm of RF and microwave engineering, precise and efficient machining techniques are critical driving forces in innovative development of technologies. As RF systems continue to evolve towards smaller footprints and higher levels of integration, the demand for precision manufacturing techniques becomes increasingly paramount. Traditional PCB (printed circuit board) techniques have been long served as the manufacturing foundation of various RF components, offering advantages in ease of fabrication and cost-effectiveness. However, PCB technology encounters numerous challenges that increasingly limit its manufacturing capacity to fully satisfy the evolving demands of modern electronic devices. These challenges encompass restricted design complexity, constrained material options, diminished precision in crafting fine features, and inefficiencies in rapid prototyping and waste management. As electronic devices progressively diminish in size while their functionality expands, the imperative for enhanced precision, superior design flexibility, and a wider array of material integration intensifies. This context accentuates the critical need for advanced manufacturing techniques capable of overcoming these challenges. In the realms of micromachining techniques, 3D printing technology, photolithography, and thin film growth methodologies are innovative methods poised to transform the fabrication of electronic components and devices. These avant-garde solutions offer a counter to the intrinsic limitations of conventional PCB technology, facilitating the creation of devices with higher resolution, intricate three-dimensional structures, swift design iterations, and an expanded spectrum of materials. The shift towards these sophisticated techniques marks a significant transformation, ready to tackle the progressive challenges of electronics manufacturing. This chapter ventures into the advanced manufacturing realms specifically designed for RF and microwave applications. Focusing on micromachining techniques, 3D printing technology, photolithography, and materials-based thin film growth methods, this discourse promises to extend the frontiers of RF and microwave device fabrication. By enabling the production of compact, high-performance components vital for contemporary and future wireless communications and radar technologies, these forefront methods are set to redefine the fabrication landscapes of RF and microwave devices. Through an in-depth examination of these technologies, this chapter endeavors to shed light on the shifting paradigms of RF manufacturing and its consequential effects on future RF devices and systems.

3.1 Micromachining Techniques

Micromachining techniques refer to a collection of processes used to fabricate extremely small parts and features, often at the microscale or even nanoscale, with high precision and accuracy. These methods have become fundamental in various fields, such as microelectronics, and micro electromechanical systems (MEMS). The ability to work at such small scales opens new possibilities for device functionality and integration, allowing for the development of highly compact and complex systems.

3.1.1 Laser micromachining

Laser micromachining is a highly precise manufacturing process that utilizes laser technology to remove material from a workpiece/sample at a micron-scale level. It's widely employed across various industries, including electronics, medical devices, aerospace, and automotive, due to its ability to produce intricate features with exceptional accuracy and minimal heat-affected zones. As illustrated in Figure 3.1, the setup for laser micromachining typically involves a laser source, optical components for beam shaping and focusing, a workpiece positioning system, and a control system. The laser source generates a high-energy beam, which is then directed through lenses, mirrors, and other optical elements to control its intensity, focus, and spot size. The workpiece is positioned precisely using a stage or gantry system, allowing for accurate positioning and movement during machining. Control software coordinates the entire process, adjusting parameters such as laser power, pulse duration, scanning speed, and beam position to achieve the desired machining results.

The focused laser beam directed onto the surface of the workpiece interacts with the material, leading to various physical processes, as depicted in Figure 3.2 [1]. When the laser beam interacts with the surface of the workpiece, the focused laser beam delivers energy to the material, leading to localized

Figure 3.1: Diagram of a laser micromachining setup.

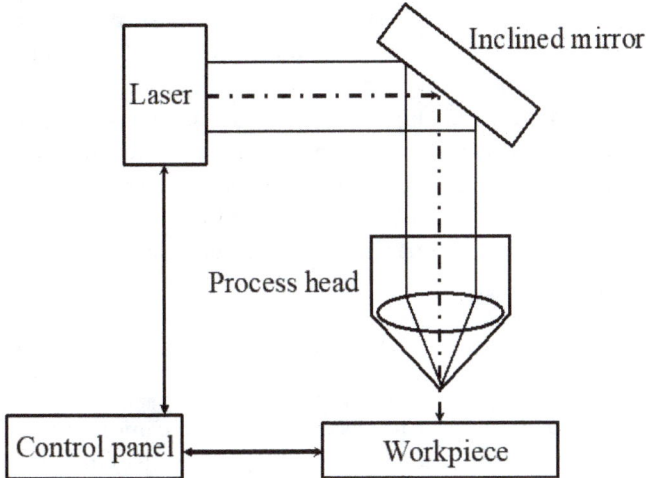

Figure 3.2: General physical process of laser micromachining: (a) heating, (b) surface melting, (c) surface vaporization, (d) plasma formation, (e) ablation.

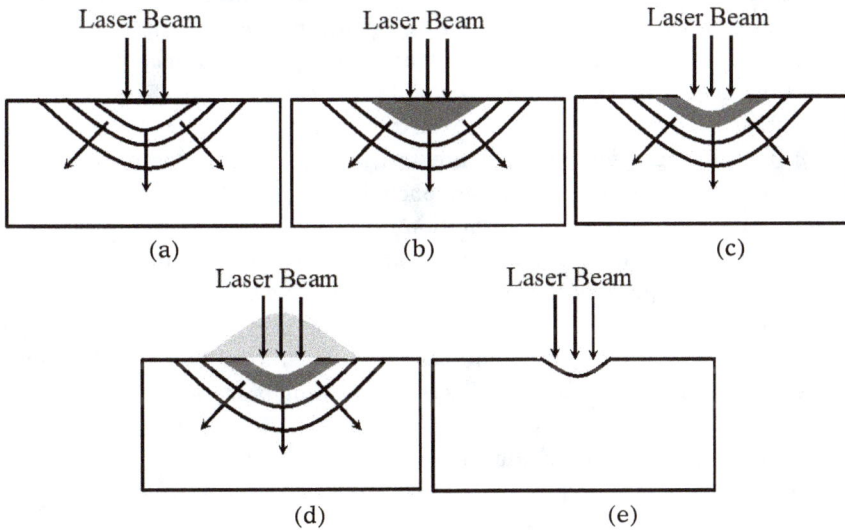

heating. This increased temperature changes the material's properties, such as softening or expansion. In areas where the laser intensity is sufficient, surface melting may occur, resulting in the formation of a molten pool on the surface. Additionally, if the energy delivered by the laser beam surpasses the material's vaporization threshold, surface vaporization takes place. This process involves the direct transition of material from solid to gas phase without passing through the liquid phase. At extremely high laser intensities, the material may undergo ionization, leading to plasma formation. Plasma, composed of ions and free electrons, is highly energized and can interact with the laser beam and the surrounding environment. Ultimately, the combination of these processes results in material removal from the surface, known as ablation.

3.1.2 Ion beam micromachining

Ion beam micromachining is a precise machining technique used to remove material from a workpiece at the microscale level. It involves directing a focused ion beam onto the surface of a workpiece, where the ions interact with the material to cause sputtering or etching, thereby removing material in a controlled manner. Ion beam micromachining offers several advantages, including high precision, sub-micron resolution, and the ability to machine a wide range of materials, including metals, semiconductors, and insulators.

Figure 3.3 illustrates the ion beam micromachining setup, which comprises ion sources capable of generating intense beams with the appropriate energy spread to effectively extract atoms from the workpiece through ion impingement [2]. In this setup, a cathode, typically a heated tungsten (W) filament, is employed to accelerate electrons using a high voltage (around 1 kV), facilitating their passage into the anode. As the electrons traverse from the cathode to the anode, they encounter argon (Ar) atoms within the plasma source, leading to the formation of Ar ions. This process induces an electron arc due to the magnetic field generated between the cathode and the anode. Subsequently, the ions produced are extracted from the plasma and directed towards the workpiece. The workpiece is positioned on a water-cooled table, which not only stabilizes the temperature but also facilitates the removal of heat generated during the machining process. Additionally, the workpiece is often tilted at an angle ranging from 0° to 180° to enable precise control over the ion beam incidence angle, thereby optimizing the machining results.

Figure 3.3: Ion beam micromachining system.

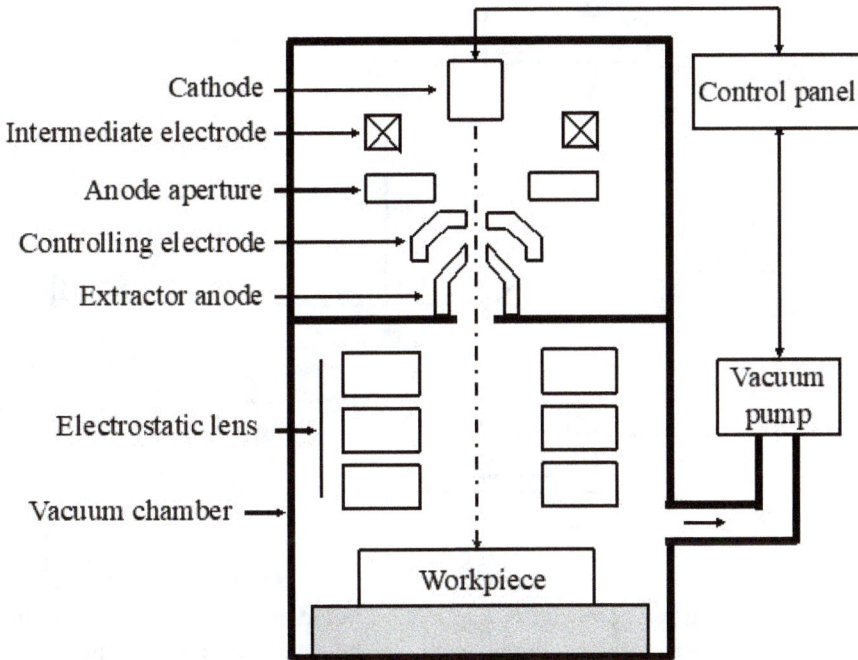

3.1.3 Micro-electro discharge machining

Micro-electro discharge machining (Micro-EDM) is an advanced machining process used to fabricate intricate shapes and features on conductive materials at the micro-scale level. This technique utilizes electrical discharges, or sparks, to erode material from the workpiece through a series of controlled electrical pulses. It is particularly well-suited for machining hard and brittle materials that are difficult to process using conventional methods, such as hardened steel, tungsten carbide, and ceramic materials. Applications of Micro-EDM include the fabrication of micro-molds, micro-structured surfaces, micro tools, micro-electrodes, etc.

As shown in Figure 3.4, the Micro-EDM process involves a tool electrode, and a workpiece submerged in a dielectric fluid, typically deionized water or

Figure 3.4: General setup of micro-electro discharge machining.

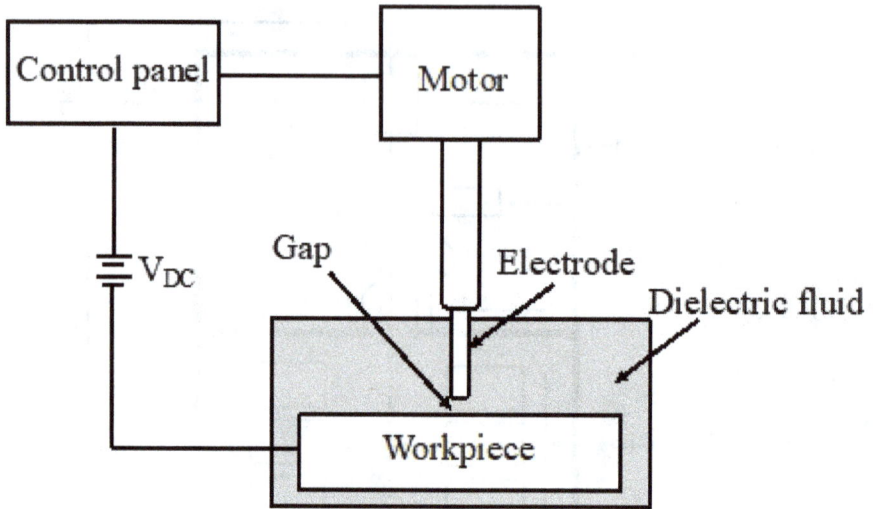

oil. When a voltage potential is applied between the tool electrode and the workpiece, an electrical discharge occurs in the small gap between them. This discharge generates a high-energy plasma channel, vaporizing and melting a tiny portion of the workpiece material. The resulting debris is flushed away by the dielectric fluid, allowing for continuous material removal. The working principle of Micro-EDM involves several sequential stages, as illustrated in Figure 3.5 [3]. Initially, high-frequency electrical pulses are applied between the electrode and the workpiece, both submerged in a dielectric fluid. This voltage differential triggers the initiation of an electric discharge. As the voltage surpasses a critical threshold, the dielectric fluid in the gap ionizes, releasing electrons that accelerate, creating an electron avalanche. This cascade of electrons generates intense heat upon collision with atoms in the dielectric fluid and workpiece material, leading to localized melting and evaporation of the workpiece material at the contact point with the electrode. The rapid heating and vaporization produce a localized pressure wave, which exerts a force on the molten material, expelling it from the workpiece surface. This erosive action results in material removal from the workpiece, contributing to the machining process. Over repeated pulses, this process forms a crater on the workpiece surface, representing the material removed during Micro-EDM.

Figure 3.5: Working principle of Micro-EDM: (a) initiation of electric discharge, (b) ionization and electron avalanche, (c) melting and evaporation, (d) erosion, (e) crater formation.

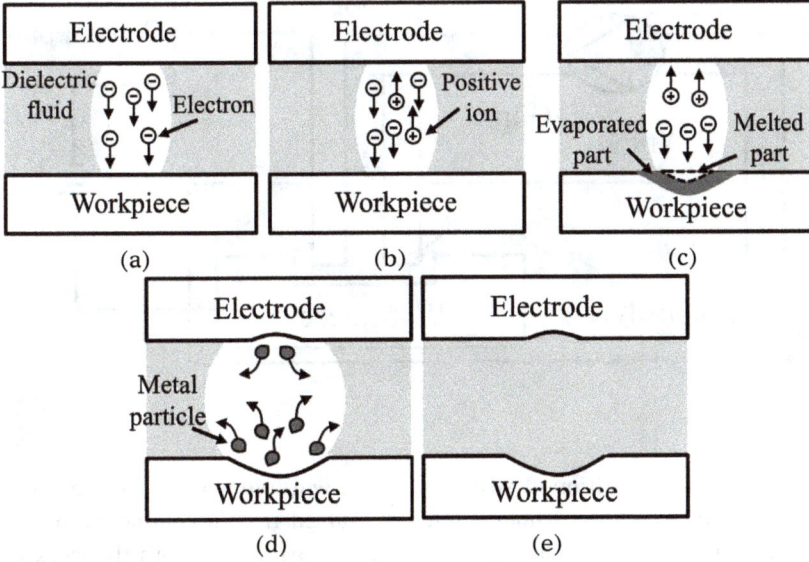

3.1.4 Electrochemical micromachining

Electrochemical micromachining (ECMM) is a precise method employed in manufacturing to delicately remove material from electrically conductive workpieces at the micro-scale level. It operates based on the principles of electrochemistry, utilizing an electrolyte and an electric potential to dissolve material from the workpiece surface. ECMM offers several advantages over traditional machining techniques, including the ability to machine complex geometries with high precision and accuracy, even in hard-to-machine materials such as titanium and nickel-based alloys. Additionally, ECMM produces minimal residual stress and heat-affected zones, making it suitable for machining delicate and heat-sensitive materials.

Figure 3.6 illustrates the ECMM setup, which comprises several key components and steps to facilitate precise material removal from the workpiece surface [4]. Initially, the workpiece, typically composed of an electrically conductive material such as metal or alloy, is securely mounted within the

Figure 3.6: Diagram of electrochemical micromachining.

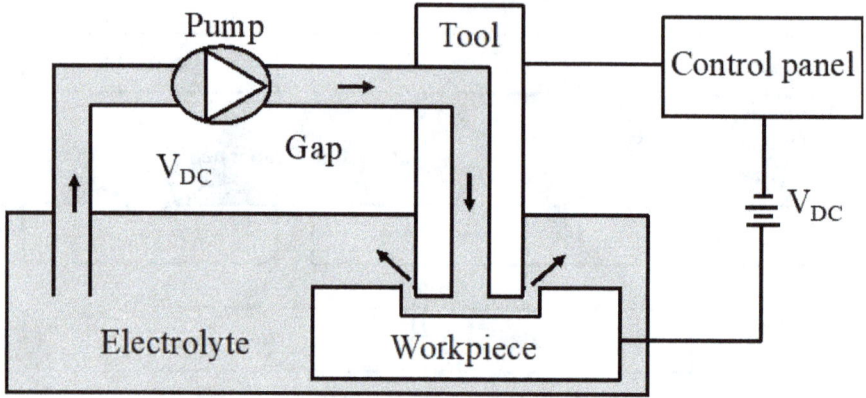

machining setup. Positioned adjacent to the workpiece is the tool electrode, also referred to as the cathode, which is crafted from a conductive material and shaped according to the desired machining geometry. Both the workpiece and the tool electrode are submerged in an electrolyte solution, serving as a medium for the electrochemical reactions to occur and facilitating material removal. A power supply is then connected to provide the necessary electrical potential between the workpiece (anode) and the tool electrode (cathode), with control over parameters like voltage, current, and polarity. This setup is governed by a control system, ensuring precise regulation of machining parameters such as voltage, current density, electrolyte flow rate, and machining duration, thereby enabling automation and precise control over the machining process. Additionally, the pump is responsible for circulating the electrolyte fluid, helping to maintain stable operating temperatures and prevent overheating of the workpiece and tool electrode.

3.2 3D Printing Technologies

3D printing technology, or additive manufacturing, is a transformative approach to industrial production that enables the creation of complex, high-precision objects by adding material layer by layer from a digital model [5]. Unlike traditional subtractive manufacturing methods, 3D printing allows for unparalleled design flexibility, rapid prototyping, and efficient material usage, making it ideal for a wide range of applications across industries. This technology encompasses a range of printing techniques, such as fused deposition modeling (FDM), stereolithography (SLA), selective laser sintering (SLS), and digital light processing (DLP), among others. Each method provides distinct advantages, facilitating the creation of tailored, detailed designs while minimizing waste and accelerating production timelines.

3.2.1 Fused deposition modeling

Fused deposition modeling (FDM) is a popular 3D printing technique that works by extruding a thermoplastic filament through a heated nozzle, which melts the material and deposits it layer by layer onto a build platform, as depicted in Figure 3.7. As the molten plastic is extruded, it quickly solidifies, forming the desired shape. The process begins with a digital 3D model sliced into thin cross-sectional layers by specialized software. The sliced model is then sent to the 3D printer, where the printing process begins. The printer's nozzle moves along the X and Y axes, while the build platform moves along the Z axis, allowing the layers of molten plastic to be deposited one on top of the other, gradually building up the final object.

Figure 3.7: Setup of FDM 3D printing.

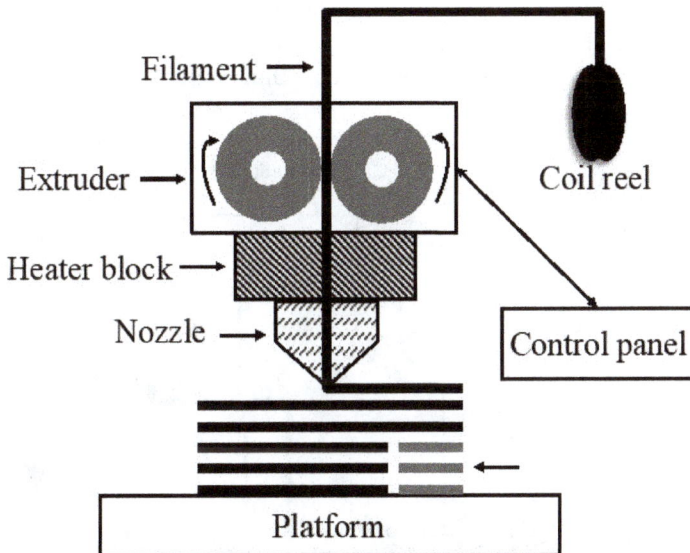

FDM boasts numerous advantages, notably its ease of use, cost-effectiveness, material diversity, and rapid prototyping capabilities. Additionally, FDM printers facilitate the creation of support structures for complex geometries, which can be effortlessly removed post-printing. Despite its renowned simplicity and versatility, FDM does present some limitations, such as comparatively lower resolution and restricted compatibility with certain materials. Nonetheless, continuous advancements in FDM technology are progressively overcoming these constraints, positioning it as an increasingly favored option across various prototyping and manufacturing applications.

3.2.2 Stereolithography

Stereolithography (SLA) is an advanced 3D printing technology widely used in various industries for creating highly detailed, intricate, and high-resolution parts and prototypes. SLA is commonly used in industries for rapid prototyping, product development, and low-volume manufacturing. Its ability to create complex geometries with fine features makes it a valuable tool for engineers, designers, and manufacturers looking to produce high-quality prototypes and functional parts efficiently.

Figure 3.8 shows a typical SLA system, where a liquid photopolymer resin is selectively cured by an ultraviolet (UV) laser beam layer by layer, solidifying the resin to form the desired 3D object. The process begins with the creation of a digital 3D model, which is then sliced into thin cross-sectional layers. These

Figure 3.8: Typical stereolithography system.

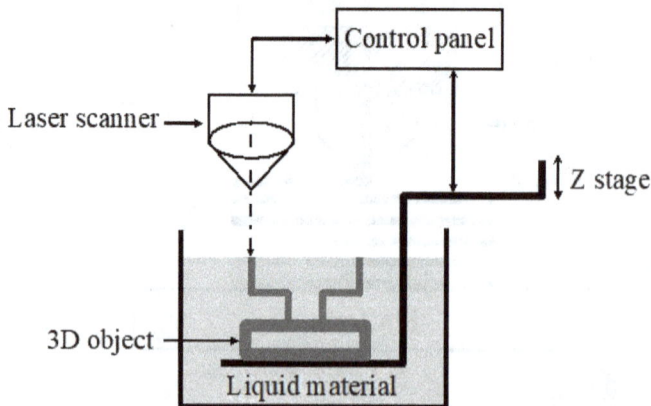

layers are sequentially exposed to the UV laser on the surface of a liquid resin vat, causing the resin to solidify where the laser beam strikes. As each layer solidifies, the build platform moves down incrementally, allowing the next layer of resin to be exposed and cured. This additive manufacturing technique enables the production of parts with intricate details, smooth surface finishes, and high dimensional accuracy.

3.2.3 Selective laser sintering

Selective laser sintering (SLS) is an additive manufacturing technology that utilizes a high-power laser to selectively fuse powdered materials, typically thermoplastics or metals, layer by layer, to create three-dimensional objects. SLS offers several advantages, including the ability to produce complex geometries without the need for support structures, as the surrounding powder acts as a self-supporting material during printing. Additionally, SLS can work with a wide range of materials, including thermoplastics, metals, ceramics, and composites, making it suitable for various industrial applications, including prototyping, tooling, and end-use part production.

In the setup of an SLS system in Figure 3.9, a thin layer of powdered material is evenly spread over the build platform, and a high-power laser is then

Figure 3.9: Typical configuration of a selective laser sintering setup.

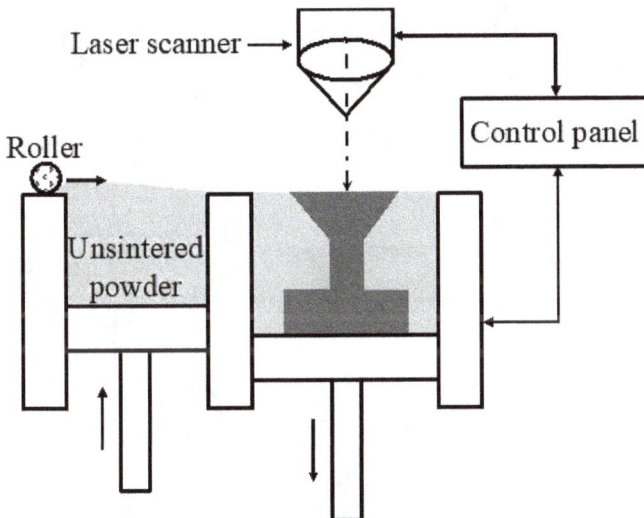

directed onto specific areas of the powder bed, selectively sintering or melting the particles together based on the digital model of the object being printed. After each layer is sintered, the build platform is lowered, and a new layer of powder is spread over the previous one. This process is repeated until the entire object is fabricated. The process flow of SLS typically involves the preparation of powder, layering, selective sintering, platform lowering, powder re-coating, and cooling and removal stages.

3.2.4 Digital light processing

Digital light processing (DLP) is a 3D printing technology that utilizes a digital light projector to selectively cure photopolymer resin layer by layer, creating three-dimensional objects. DLP offers several advantages, including fast printing speeds, high resolution, and the ability to produce intricate details and smooth surface finishes. Additionally, DLP printers can accommodate a wide range of photopolymer materials, offering versatility in material properties and colors.

As shown in Figure 3.10, a vat of liquid resin is exposed to light from a digital projector, which contains a micro-mirror array that reflects light onto the resin surface. The projected light cures the resin in specific areas, solidifying it to form each layer of the object being printed. As each layer is cured, the build platform gradually moves upwards, allowing the next layer of resin to be

Figure 3.10: Schematic of a digital light processing 3D printer.

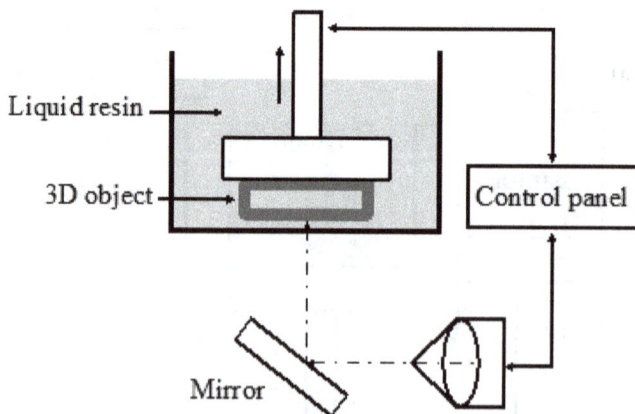

exposed and cured. This additive manufacturing process results in the gradual construction of the desired object with high precision and resolution.

3.3 Photolithography Techniques

Photolithography, a cornerstone technique in microfabrication and semiconductor manufacturing, harnesses light to transfer geometric patterns from a mask to a light-sensitive chemical photoresist layer on a substrate. Photolithography's precision and ability to produce extremely fine patterns have made it essential in the production of integrated circuits and microelectromechanical systems. Advancements in photolithography, including the development of deep ultraviolet and extreme ultraviolet lithography, continue to push the boundaries of miniaturization, enabling the creation of features at the nanometer scale. The technique's adaptability to various materials and its capability to create highly complex patterns with high throughput make it a fundamental technology in modern electronics and microfabrication.

Lithography can be broadly categorized into two primary methods: imaging techniques and direct writing, as outlined in Table 3.1. Imaging techniques involve the use of a lithographic mask that transfers patterns to a substrate coated with a light-sensitive resist, typically through photon imaging. Conversely, direct writing eschews the use of masks, instead employing a focused beam of electrons, ions, or atoms that is directly scanned across the substrate, allowing for precise pattern creation.

Table 3.1: Standard lithography techniques

Imaging techniques	Photolithography (visible light)
	X-ray lithography
	Ultraviolet or extreme ultraviolet lithography
Direct writing	Electron beam lithography
	Ion beam lithography
	Atom beam lithography

3.3.1 Imaging techniques

The photolithography process, particularly when employing imaging techniques, involves several critical steps to transfer intricate patterns from a photomask onto a substrate [6]. Figure 3.11 shows the general process of imaging techniques using a positive photoresist. Initially, the substrate is coated with a uniform layer of photoresist, a light-sensitive polymer. This coating process is typically followed by a soft bake (up to 30 min at temperatures between 60 and 100 oC) to evaporate any solvents, solidifying the photoresist in preparation for exposure. The substrate is then positioned beneath a photomask, which contains the desired pattern, and exposed to ultraviolet light. The exposure initiates a chemical transformation within the positive photoresist, making the exposed areas soluble.

Figure 3.11: Schematic representation of photolithography using positive resist.

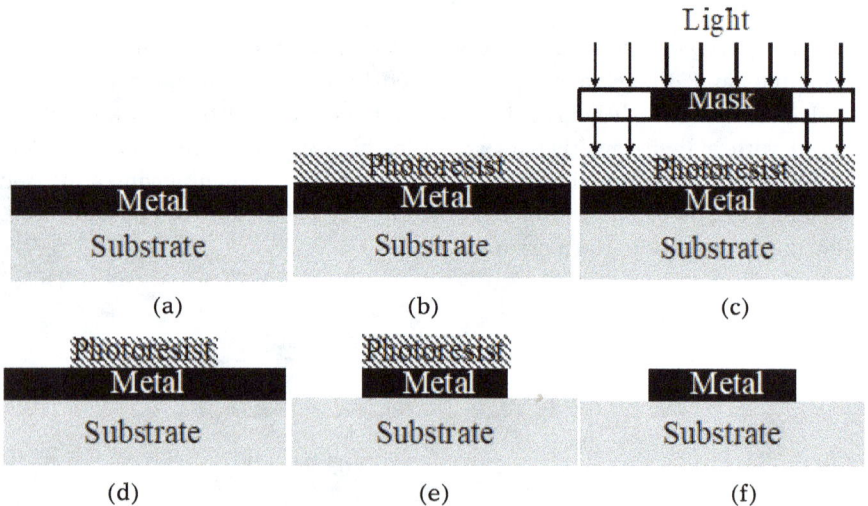

After exposure, the substrate undergoes a post-exposure bake (20–30 min at temperatures between 120 and 180°C) to further develop the photoresist's photochemical reactions, enhancing the resolution of the pattern. The development stage follows, where the substrate is bathed in a developer solution, removing the exposed or unexposed photoresist depends on the polarity, revealing the underlying substrate patterned according to the photomask. Finally, the developed substrate is etched if needed, transferring the pattern onto the substrate material. The etching can be done using various techniques, including wet chemical etching or dry plasma etching, depending on the desired precision and material properties. Once etching is complete, the remaining photoresist is stripped away with solvent or photoresist stripper, leaving behind the desired patterned features on the substrate. This meticulous process allows for the fabrication of highly complex and precise microscale and nanoscale structures.

Conversely, when using a negative photoresist, the process is similar, but with a key difference in the behavior of the photoresist upon exposure to UV light. In this case, depicted in Figure 3.12, the exposed areas of the negative photoresist become insoluble to the developer solution, whereas the unexposed regions are washed away. This inversion of solubility means that the pattern on the photomask is transferred as a negative image to the substrate, allowing for the creation of structures where the exposed photoresist remains on the surface.

Figure 3.12: Schematic of photolithography using negative resist.

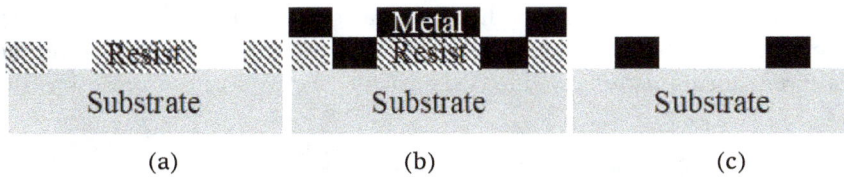

(a)   (b)   (c)

The primary benefit of imaging technique-based photolithography lies on its capacity for parallel processing, which is crucial for mass production. Nonetheless, its effectiveness is constrained by the Abbe–Rayleigh criterion:

$$d_{min} = 0.61 \frac{\lambda}{n \sin\alpha} = 0.61 \frac{\lambda}{NA} \tag{3.1}$$

where d_{min} is the minimal size of a structure illuminated with a point light source with wavelength λ, n is the refraction index of the environment, α is the aperture angle, and NA is the numerical aperture of the system. Therefore, to fabricate small structures, it's necessary to utilize high numerical apertures and shorter wavelengths. For structures larger than 250 nm, a high-pressure mercury lamp suffices. However, for features between 150 and 250 nm, shorter wavelengths are essential. To achieve dimensions smaller than 100 nm, light wavelengths of less than 150 nm are required, a feat achievable through extreme ultraviolet lithography, which operates at wavelengths as low as 13.4 nm.

3.3.2 Direct writing

Direct writing photolithography represents a highly precise approach to pattern fabrication, distinguished by its method of selectively removing or adding material on a substrate without the use of a physical mask. This technique involves the direct application of focused beams of electrons, ions, or photons to a surface coated with a sensitive resist, enabling the creation of intricate patterns with extremely high resolution. Direct writing photolithography's high precision and ability to directly create complex patterns make it ideal for prototyping and small-scale production, despite its lower throughput compared to traditional photomask-based methods. However, the technique's relatively slow speed compared to mask-based methods can limit its use in high-volume manufacturing contexts.

As shown in Figure 3.13, the general process of direct writing photolithography commences with meticulous substrate preparation, which may include the deposition of a metal layer to serve as the foundation for subsequent patterning. Following this, a resist layer sensitive to the writing beam (electrons, ions, or photons) is applied through methods such as spin coating to ensure uniform coverage. The substrate undergoes a soft bake to evaporate solvent from the resist, enhancing its adhesion and sensitivity. The core of the process is the

Figure 3.13: General process of direct writing photolithography.

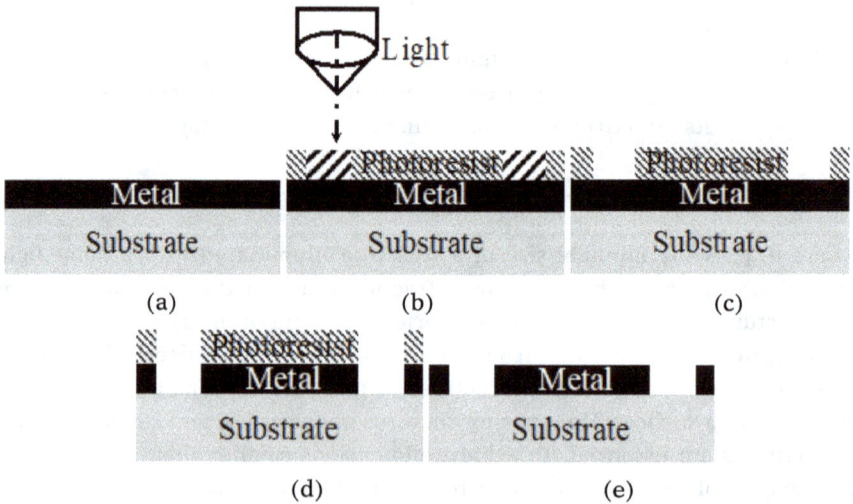

exposure phase, where a focused beam, controlled by a pattern generator, selectively modifies the resist's solubility properties according to the desired pattern. After exposure, the substrate is developed in a solution that dissolves either the exposed areas of a positive resist or the unexposed areas of a negative resist, creating the pattern. A post-exposure bake may follow to harden the pattern, which then undergoes further processing steps like etching or lift-off for pattern transfer. The process concludes with cleaning to remove any remaining resist, leaving behind the desired patterned structure on the substrate.

3.4 Thin Film Growth Methods

Thin film growing techniques are pivotal processes, involving the deposition of materials onto a substrate under precisely controlled conditions to form films with specific properties. These techniques allow for precise control over film thickness, composition, and properties, facilitating the miniaturization of RF and microwave devices and integration of diverse materials. With their ability to produce nanoscale or microscale thin films, they contribute to the development of advanced device architectures and materials tailored for high-frequency applications. Thin film growing techniques can be broadly categorized into four main types: physical vapor deposition (PVD), chemical vapor deposition (CVD), atomic layer deposition (ALD), and molecular beam epitaxy (MBE). Each of these thin film growth techniques has its advantages and application areas. Through these techniques, scientists and engineers can fabricate materials and structures with highly customized properties, driving advancements in numerous high-tech domains.

3.4.1 Physical vapor deposition

Physical vapor deposition (PVD) is a vacuum coating process that is widely used to deposit thin films of various materials on substrate surfaces. This technique involves the physical transfer of material from a solid or liquid source to the substrate through the phase of vapor. PVD processes are conducted under vacuum conditions and involve three main steps: vaporization of the coating material, transportation of the vapor to the substrate, and condensation of the vapor onto the substrate, forming a thin film. There are two principal methods of PVD, including sputtering, in which particles are ejected from a solid target material due to bombardment of the target by energetic particles; and evaporation, where the material is heated until it vaporizes.

Figure 3.14 (a) illustrates a general setup of a PVD sputtering system. The sputtering process initiates as the sputtering gas, typically argon, is ionized within a vacuum chamber, generating ions (e.g., argon ions, Ar^+). These ions are then accelerated towards the target material, which bears a negative charge. Upon impact, the kinetic energy of the ions is transferred to atoms of the target, causing some atoms to be knocked out of the target surface. These dislodged atoms then condense on the substrate, gradually building up as a thin film. Sputtering allows for the deposition of a wide range of materials, including metals, insulators, and compounds, with excellent control over the film's composition and thickness. It is distinguished by its ability to coat materials with high melting points and to achieve a high degree of uniformity over complex shapes and large areas.

One of the key advantages of sputtering is its flexibility; it can be adapted to various configurations, including magnetron sputtering, reactive sputtering, and co-sputtering, each offering unique capabilities for specific applications. Magnetron sputtering, for instance, utilizes magnetic fields to confine plasma close to the target, enhancing the efficiency of the sputtering process and enabling the deposition of films at lower pressures and with higher deposition rates.

Evaporation is another fundamental process in PVD used to deposit thin films onto substrates. As shown in Figure 3.14 (b), the evaporation process typically takes place within a vacuum chamber to prevent unwanted chemical reactions and ensure the purity of the deposited film. The source material can be heated using various methods, including resistive heating, electron beam

Figure 3.14: Physical vapor deposition methods: (a) sputtering, (b) evaporation.

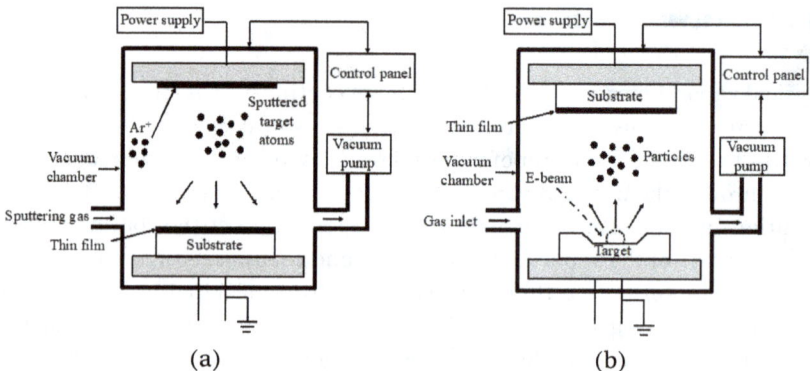

(a)   (b)

heating, or laser ablation, depending on the specific requirements of the application. During evaporation, the vaporized atoms travel in straight lines within the vacuum chamber before reaching the substrate. The film's thickness and uniformity are controlled by factors such as the deposition rate, temperature of the source material, and substrate temperature.

Evaporation is widely used in PVD processes due to its simplicity, versatility, and ability to deposit high-purity films. It is particularly suitable for depositing metals, alloys, and some compounds, making it valuable in applications such as optical coatings, semiconductor manufacturing, and thin film electronics.

3.4.2 Chemical vapor deposition

Chemical vapor deposition (CVD) is a widely used material growth technology that allows for the deposition of solid, high-purity, high-performance thin films. In a CVD process, a substrate is exposed to one or more volatile gas precursors, which react and/or decompose on the substrate surface to produce the desired deposit. Unlike PVD, where physical processes dominate, CVD relies on chemical reactions to achieve film growth. The major advantages of CVD include the ability to deposit films on complex-shaped substrates with excellent step coverage, the potential for high-purity films due to the chemical purification effects during deposition, and the capability to deposit a wide range of materials, including metals, semiconductors, dielectrics, and polymers.

CVD processes can be categorized into several types based on the energy source used to activate the precursor gases, including thermal CVD, plasma-enhanced CVD (PECVD), and laser-assisted CVD (LCVD), among others. The choice of process depends on the materials to be deposited, the desired properties of the film, and specific application requirements. Figure 3.15 shows a general setup of thermal CVD. This method involves heating a substrate to high temperatures, typically between 600 °C and 1000 °C, in a reaction chamber. Volatile precursor gases are introduced into the chamber, where they react or decompose upon contacting the heated substrate, resulting in the deposition of a solid material layer. The entire process hinges on the precise control of temperature, precursor gas flow rates, and chamber pressure to achieve the desired film thickness and material properties. As the reaction proceeds, the substrate is uniformly coated, layer by layer, with a high-quality film that exhibits excellent adhesion and uniformity. Excess gases and by-products of the reaction are continuously removed by the vacuum system, ensuring a clean deposition environment. The thermal CVD process is renowned for its ability to produce films with superior purity and conformity, making it indispensable for creating complex structures in electronic and RF device manufacturing.

Figure 3.15: General setup of thermal CVD.

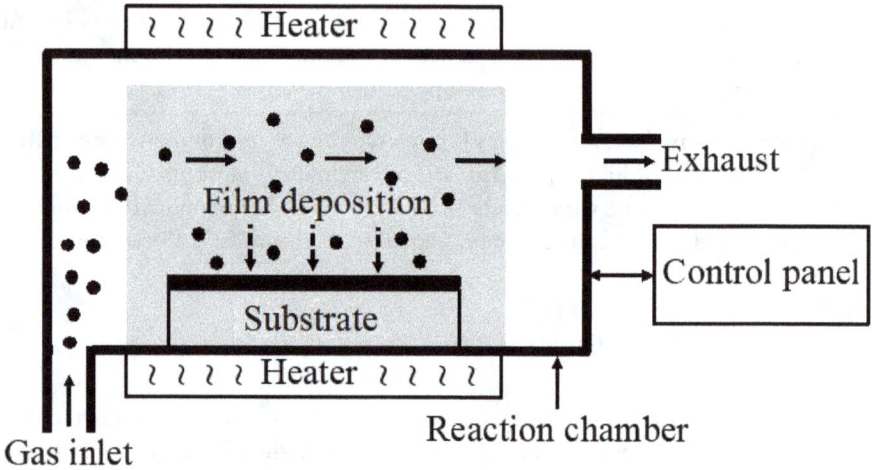

3.4.3 Atomic layer deposition

Atomic layer deposition (ALD) is a vapor-phase technique used for depositing thin films onto substrates with exceptional control over thickness, composition, and conformality. ALD can be considered a subset of CVD, as both share similar setup configurations and operational principles. However, ALD distinguishes itself through its unique process of depositing thin films via sequential, self-limiting surface reactions. Each cycle of the process adds a single atomic layer to the surface, allowing for precise control over the film growth. ALD's ability to coat complex three-dimensional structures uniformly, even inside deep trenches or around high-aspect-ratio features, makes it particularly valuable for advanced semiconductor fabrication, where it is used to create high-dielectric-constant dielectrics, metal gates, and other critical components. Additionally, its low processing temperature makes it suitable for a wide range of materials and applications

It operates through a cyclical process involving the alternate pulsing of chemical precursors and inert gas purges into a reaction chamber. Figure 3.16 shows the general process of atomic layer deposition. Initially, a precursor is introduced and chemically bonds to the substrate in a self-limiting fashion, ensuring a monolayer deposition. This step is followed by an inert gas purge to clear the chamber of excess precursors and reaction byproducts. Subsequently, a second precursor is pulsed in, reacting with the first layer to form the desired material, with another purge cycle ensuring the removal of any residuals. This meticulous, layer-by-layer approach allows for exceptional control over film properties, making ALD invaluable for applications requiring uniform coatings on intricate 3D structures, such as in advanced semiconductor manufacturing.

Figure 3.16: Atomic layer deposition process.

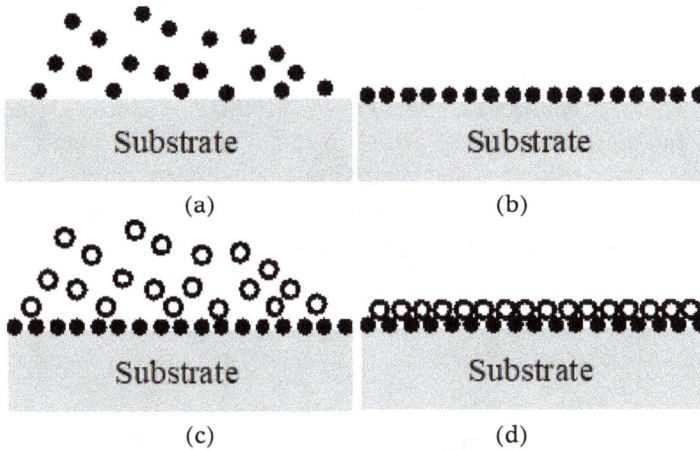

(a)

(b)

(c)

(d)

3.4.4 Molecular beam epitaxy

Molecular beam epitaxy (MBE) is a highly controlled thin-film deposition technique used in the fabrication of semiconductor devices. It involves the slow deposition of molecules onto a substrate under ultra-high vacuum conditions, typically in the range of 10^{-8} to 10^{-11} Torr. The process allows for the precise layer-by-layer growth of crystalline materials. One of the key advantages of MBE is its exceptional control over the thickness and composition of the layers, down to a single atomic layer. This precision enables the production of high-quality quantum wells, superlattices, and other nanostructures with specific electronic and optical properties. The technique also allows for real-time monitoring and control of the growth process through various in-situ diagnostic tools like reflection high-energy electron diffraction (RHEED), which provides information on the surface structure during deposition. MBE is crucial in the research and development of semiconductor materials such as GaAs, InP, and other III–V and II–VI compounds, making it indispensable for the advancement of electronics and optoelectronics, including lasers, light-emitting diodes (LEDs), and high-speed transistors. Its ability to create high-purity, defect-free materials with precise control over composition and doping levels makes it a preferred method for fabricating devices requiring atomic-scale engineering.

MBE is performed in an ultra-high vacuum chamber, as illustrated in Figure 3.17, to minimize contamination and ensure the accurate deposition of material layers. Inside this chamber, effusion cells, also known as Knudsen cells, contain the source materials. These cells are heated to vaporize the material, creating a beam of atoms or molecules that travel toward the substrate. A substrate holder is precisely controlled for orientation and temperature, ensuring optimal conditions for film growth. Mechanical shutters strategically regulate the molecular beams' access to the substrate, allowing for precise control over layer thickness and composition. Monitoring equipment, such as RHEED, provides real-time feedback on the growth process, enabling adjustments to be made on the fly to ensure the desired material properties are achieved.

Figure 3.17: Schematic representation of a molecular beam epitaxy growth chamber.

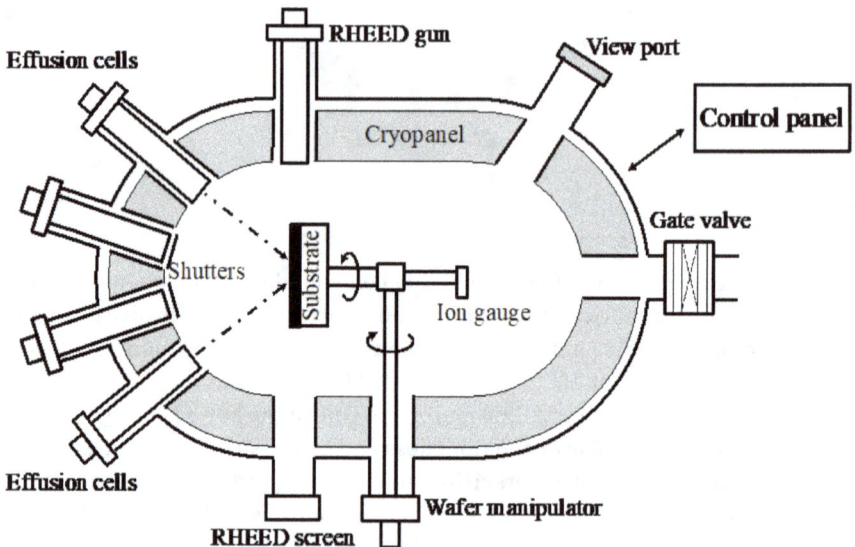

The MBE fabrication process starts with the meticulous preparation of the substrate, which is cleaned and placed in the substrate holder within the vacuum chamber. The substrate is then heated to a predetermined temperature that promotes optimal material adhesion and crystalline structure. As the effusion cells vaporize the source material, the resulting molecular beam is directed towards the substrate, where it condenses to form thin, uniform layers. The use of shutters allows for the deposition process to be carefully controlled, enabling the construction of complex multilayer structures with precise compositional gradients. Throughout the deposition, the growth of the thin film is closely monitored using RHEED, which provides valuable information on the surface structure and growth rate of the material. This allows for real-time adjustments to the process parameters, ensuring the epitaxial layers meet the desired specifications. Once the deposition is complete, the substrate is cooled and removed from the chamber for further processing or analysis.

In summary, the exploration of novel fabrication and manufacturing techniques, including micromachining, 3D printing, photolithography, and thin film growth methods, demonstrates the significant advancements in the precision, scalability, and versatility of RF and microwave component production. These techniques not only enable the creation of complex structures with high resolution but also open new possibilities for integrating advanced materials and achieving tailored functionalities. As the demand for more sophisticated and miniaturized devices continues to grow, these innovative manufacturing approaches will play a critical role in meeting the evolving needs of modern technology.

References

[1]   Samant A N, Dahotre N B., Laser machining of structural ceramics—A review. *Journal of the European Ceramic Society*, 29(6), 969-993, 2009.

[2]   Daud N D, Hasan M N, Saleh T, et al., Non-traditional machining techniques for silicon wafers. *The International Journal of Advanced Manufacturing Technology*, 121(1): 29-57, 2022.

[3]   Takahata, Kenichi, *Micro electronic and mechanical systems*. BoD–Books on Demand, 2009.

[4]   Bhargav, K. V. J., et al., Experimental investigation on machining characteristics of titanium processed using electrolyte sonicated Â-ECDM system. *Scientific Reports*, 12(1), 15540, 2022.

[5]   Sole-Gras, M., Ren, B., Ryder, B.J. et al., Vapor-induced phase-separation-enabled versatile direct ink writing. *Nature Communications*, 15, 3058, 2024.

[6]   J. Ge, R. Floyd, A. Khan and G. Wang, High-Performance Interconnects With Reduced Far-End Crosstalk for High-Speed ICs and Communication Systems. *IEEE Transactions on Components, Packaging and Manufacturing Technology*, 13(7), 1013-1020, 2023.

4

Reconfigurable RF and Microwave Technologies

With the advent of high-end manufacturing processes, RF technology stands at the cusp of a transformative era. Techniques, ranging from precision machining to additive manufacturing, have not only enhanced the fabrication of RF and microwave components but also expanded the possibilities for intricate designs and functionalities. As a result, RF systems can now be tailored with greater precision to meet the exacting demands of modern applications, spanning from wireless communication networks to radar systems and beyond. Moreover, amidst this technological evolution, the significance of tunable and reconfigurable RF technologies emerges prominently. These methodologies empower RF systems with adaptability and flexibility, enabling them to dynamically adjust their operating parameters in response to changing environmental conditions, signal requirements, and system constraints. By incorporating tunable and reconfigurable elements into RF designs, engineers can optimize performance, enhance spectrum efficiency, and streamline system integration, thereby unlocking new avenues for innovation and advancement in RF engineering.

Significant efforts have been dedicated to the development of reconfigurable RF components, utilizing technologies such as mechanical tuning, semiconductor varactors, microelectromechanical systems (MEMS), and thin film technologies including ferroelectric and ferromagnetic films. Mechanical tuning involves physically altering the dimensions or configurations of RF components to modify their electrical properties, such as resonance frequency or impedance. Semiconductor varactors, on the other hand, utilize the voltage-dependent junction capacitance of semiconductor devices to achieve frequency agility and tuning in RF circuits. Microelectromechanical systems (MEMS) integrate mechanical elements and electronics on a single chip, enabling the creation of reconfigurable components like switches, tunable capacitors, and resonators. Finally, thin film technologies, particularly those leveraging ferroelectric and ferromagnetic materials, provide compact and low-power solutions for RF reconfigurability. Ferroelectric films enable voltage-controlled capacitors with tunable dielectric constants, while ferromagnetic films facilitate tunable inductors and magnetic devices, offering versatile options for automatized RF circuits. The performance, advantages, and limitations of major tuning technologies are summarized in Table 4.1.

Table 4.1: Comparison of different tuning technologies

Methods	Mechanical tuning	Varactor	MEMS	Ferroelectrics	Ferromagnetics
Q	>1000	30–50	50–400	30–150	>500
Tuning Speed	>10 µs	ns	µs	ns	ns
Bias	>1100 V	<30 V	20–100 V	<30 V	50–250 Oe
Linearity (IIP3: dBm)	High	10–35	>60	10–35	<30
Power handling	High	~mW	1–2 W	~mW	2 W
Power usage	High	Low	Very low	Very low	High
Size	Large	Small	Small	Small	Large
Cost	High	Low	Medium	Low	High
Integration	Difficult	Good	Good	Good	Difficult

4.1 Mechanical Tuning Methods

Mechanical tuning methods have been traditionally used in RF and microwave applications where robust, repeatable tuning is necessary. Mechanical tuning remains relevant in scenarios where electronic tuning might not provide the necessary robustness, power handling capability, or precision. While mechanical systems can be bulky and slow compared to electronic or MEMS-based solutions, their durability and reliability in harsh environments or high-power applications often make them the preferred choice. In the subsequent sections, various common design examples leveraging mechanical tuning methods will be explored, showcasing their utility in achieving reconfigurability and flexibility in RF and microwave applications.

4.1.1 Tunable capacitors and inductors

Figure 4.1 illustrates two tunable capacitors enabled by mechanical tuning methods: vacuum variable capacitors and air dielectric variable capacitors. Vacuum variable capacitors are essential components in RF and microwave circuits, renowned for their high performance and versatility. These capacitors consist of two sets of metal plates separated by a vacuum or gas-filled chamber, with one set fixed and the other movable. The capacitance can be adjusted by mechanically varying the spacing between the plates using a screw or motor-driven mechanism. This mechanical tuning allows for precise control over the capacitance value, enabling frequency agility, impedance matching, and tuning in RF circuits. Vacuum variable capacitors are favored for their low losses, high voltage handling capabilities, and excellent stability over a wide range of operating conditions, making them indispensable in applications such as RF amplifiers, transmitters, antennas, and tunable filters.

Figure 4.1: Capacitors enabled by mechanical tuning methods: (a) a vacuum variable capacitor, (b) an air dielectric variable capacitor.

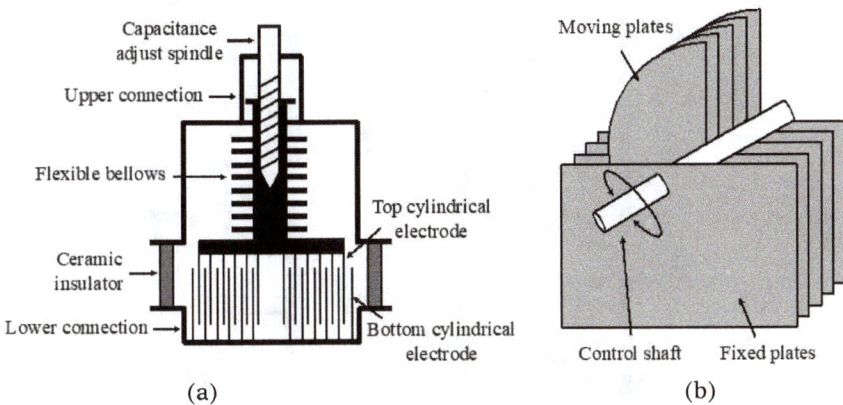

Transitioning from vacuum variable capacitors, air dielectric variable capacitors offer a more compact and cost-effective alternative while still providing reliable performance in RF and microwave circuits. These capacitors feature two sets of metal plates separated by air, rather than a vacuum or gas-filled chamber. Similar to vacuum variable capacitors, the capacitance of air dielectric capacitors can be adjusted by mechanically altering the spacing between the plates. Despite their simple constructions, air dielectric variable capacitors maintain good electrical characteristics, including low loss and high voltage handling capabilities. They have been widely used in RF tuning circuits, impedance matching networks, and variable filters, offering flexibility and efficiency in a variety of applications.

Mechanical tuning methods can also be employed to achieve tunable inductors. Figure 4.2 illustrates two common types of inductors with adjustable inductance values enabled by mechanical tuning: roller inductors and slug-tuned inductors. A roller inductor is a type of tunable inductor commonly used in RF and microwave circuits. It consists of a coil wound on a form, with a movable contact or roller that can adjust the effective length of the coil. By mechanically adjusting the position of the roller along the coil, the inductance can be varied. Roller inductors are often used in applications where precise tuning of inductance is required, such as in antenna matching networks, RF filters, and impedance matching circuits. They offer a high Q factor and low loss, making them suitable for high-frequency applications.

Figure 4.2: Inductors enabled by mechanical tuning methods: (a) a roller inductor, (b) a slug-tuned inductor.

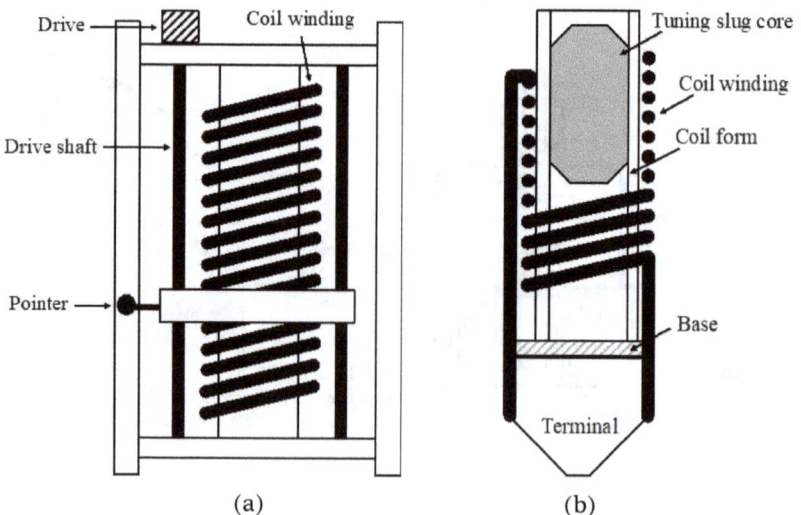

(a)   (b)

In contrast, a slug-tuned inductor, also known as a coil or choke, operates on a similar principle but with a different design. It consists of a coil wound on a form with a ferromagnetic slug inserted into the coil. By mechanically adjusting the position of the slug within the coil, the inductance can be varied. Slug-tuned inductors are widely used in applications where variable inductance is required, such as oscillator circuits, impedance matching networks, and RF amplifiers. They offer a compact design and high inductance values, making them suitable for various RF and microwave applications.

4.1.2 Tunable filters

In addition to realizing basic capacitor and inductor components, mechanical tuning methods are also applied in more intricate filter designs to achieve recon-figurable filtering performance [1]. These methods facilitate the adjustment of filter parameters such as center frequency, bandwidth, and selectivity by mechanically altering the configuration or properties of the filter components. This versatility allows for the creation of tunable filters tailored to specific frequency ranges and application requirements in various RF and microwave systems.

Figure 4.3 illustrates a rapid ($\sim$1 second) mechanical tuning system designed for high-temperature superconductor microstrip line filters [1]. Comprising high-resolution step motors, a compact pulse tube refrigerator capable of reaching a minimum temperature of 70 K, and a programmable control box, this system offers precise control over tuning parameters. Within the chamber, the tuning plate, featuring apertures for the passage of both dielectric and

Figure 4.3: Schematic of the automatic tuning system for high-temperature superconductor filters.

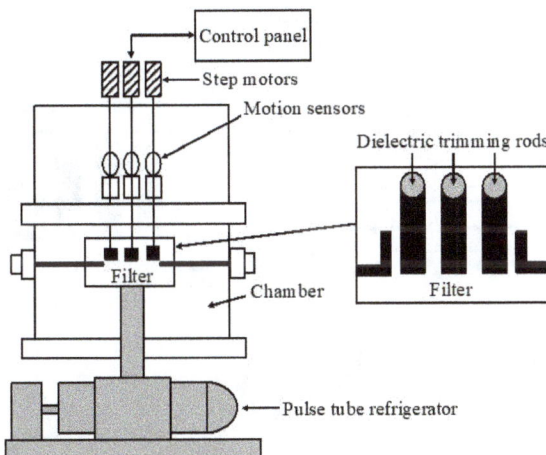

conducting trimming rods, takes center stage. Tailored for a three-pole filter operating at a 5 GHz center frequency with a 100 MHz bandwidth, this setup delivers notable performance enhancements. Utilizing both types of trimming rods, improvements in insertion loss, reduction of pass-band ripples, and bandwidth expansion are achieved. The positioning of the plate and rods above the filter is meticulously adjusted via the step motors. Remarkably, the center frequency can be shifted by over 400 MHz utilizing the tuning plate with a dielectric constant of 45. Meanwhile, bandwidth preservation is largely upheld through adjustments in couplings using copper rods. In summary, this automatic tuning system stands out as an efficient and reproducible mechanical tuning method for communication systems, offering promising avenues for enhanced performance and adaptability.

Another exemplary illustration of mechanical tuning filters is presented in Figure 4.4 [2]. This filter design is implemented with substrate integrated waveguide (SIW) technology, which involves a synthesized waveguide within a planar dielectric substrate featuring two rows of metallic vias. SIW filters share conceptual similarities with filters implemented through waveguide technology and offer various advantages, such as straightforward fabrication, compact size, low loss, complete shielding, and seamless integration with active devices. Traditionally, waveguide filters are tuned by incorporating screws into the resonant cavities and coupling apertures. Drawing inspiration from this method, this design incorporates a specially engineered tunable SIW resonator, which is realized through the introduction of a slot in the top layer and an additional

Figure 4.4: A tunable substrate integrated waveguide filter using tuning screws.

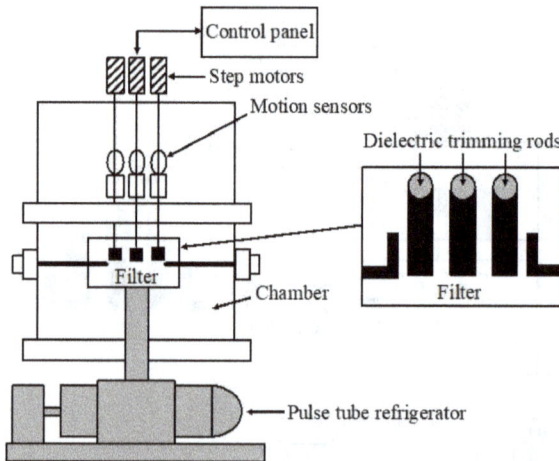

metallized via-hole in the SIW cavity. The filter depicted in Figure 4.4 offers a four-pole order filtering response centered at approximately 10 GHz. Through the adjustment of tuning screws, this filter achieves a remarkable bandwidth increment of up to 100% and exhibits a tunability of the central frequency by 10%.

In summary, mechanical tuning methods are pivotal in filter design, providing precise control over filter parameters. Techniques like tuning screws or specially designed resonators allow adjustments to the central frequency, bandwidth, and other characteristics. The integration of mechanical tuning techniques empowers engineers to enhance filter performance, meet desired specifications, and adapt to diverse operational needs.

4.1.3 Tunable antennas

Mechanical tuning methods can also find application in the design of tunable antennas. These antennas offer the flexibility to adapt to changing environmental conditions, frequency requirements, or communication scenarios. Common mechanisms for tuning include the use of mechanical actuators, switches, or movable elements such as directors, reflectors, or parasitic elements [4]. By adjusting these elements, mechanically tunable antennas can achieve alterations in parameters such as resonance frequency, impedance matching, radiation pattern, or polarization. This adaptability makes them valuable in various applications, including wireless communication systems, radar systems, satellite communications, and mobile devices, where dynamic performance optimization is essential.

Figure 4.5 showcases a frequency reconfigurable antenna, leveraging mechanical tuning methods [5]. The design integrates a slot antenna with a metasurface featuring isosceles triangular-loop unit cells. Positioned between the bottom slot antenna and the top metasurface is a dielectric substrate, with the antenna geometry adopting a circular shape for convenient rotation. A 50 Ohm coaxial cable connects to the feed point to supply RF power to the antenna, achieving a realized gain exceeding 5 dBi. Through pivoting the top metasurface with θ from 0° to 90°, the effective permittivity of the substrate can be altered, facilitating a tunable operating frequency spanning from 2.51 GHz to 3.70 GHz. Moreover, the reconfigurable antenna maintains a relatively high radiation efficiency of up to 90% across the tuning range. Overall, this design offers several advantages, including a low profile, ease of tuning, and a compact structure. These attributes render it suitable for advanced multi-standard wireless communication systems, enabling fulfillment of diverse requirements without an increase in volume.

Figure 4.5: Mechanically reconfigurable antenna: (a) top view, (b) bottom view.

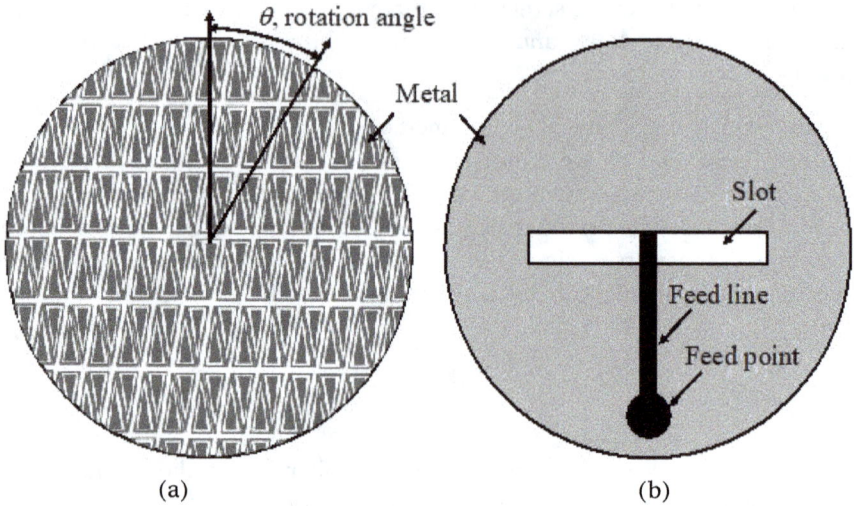

θ, rotation angle

Metal

Slot

Feed line

Feed point

(a)   (b)

Another noteworthy example of a mechanically reconfigurable antenna is depicted in Figure 4.6 [6]. This antenna is a linearly polarized reconfigurable conformal array operating at the Ku band (16–18 GHz). The array comprises an eight-way radial waveguide network loaded with eight slotted-cavity radiating elements. Symmetrically fed by a vertical 50 Ohm SMA connector, the radial waveguide network facilitates the radial propagation of a TEM mode, which is coupled to the fundamental TE_{10} mode of each rectangular waveguide output port. The symmetrical configuration of the waveguide network ensures uniform power distribution in terms of both amplitude and phase.

To adjust the antenna's performance, metallic screws are employed to modify the field distribution of the radial wave propagated inside the waveguide network. These screws establish physical contact between the upper and lower internal metallic plates of the radial waveguide network, resulting in changes to the radiating elements. Additionally, these metallic screws contribute to additional isolation of specific radiating elements within the conformal array, while maintaining an acceptable input matching bandwidth. As a result, the symmetrical placement of the screws in the radial waveguide enables main beam scanning every 45° in the azimuth plane (xy-plane), thereby yielding 65 different radiation patterns. This capability enhances the antenna's adaptability and versatility across various communication scenarios.

Figure 4.6: Illustration of the mechanically reconfigurable conformal array antenna: (a) schematic view, (b) cutting side view.

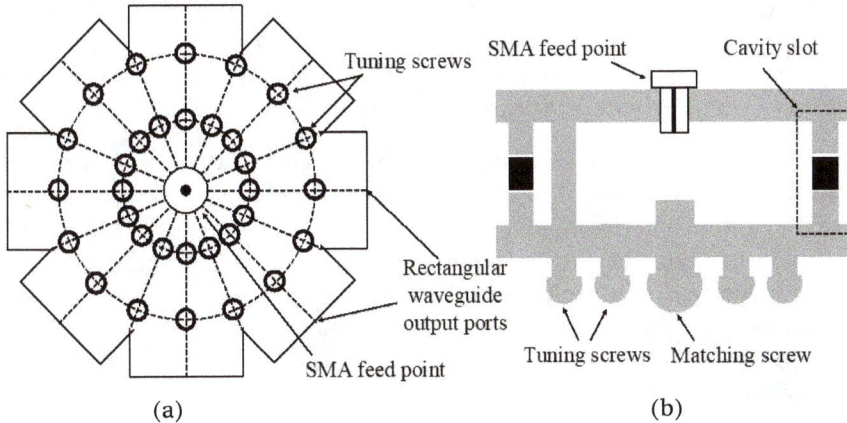

(a)   (b)

4.2 Semiconductor and RF MEMS Switches

With the continuous advancement of technology, alongside traditional mechanical tuning methods, tunable technologies based on semiconductors and MEMS have rapidly developed and found widespread applications. In comparison to traditional mechanical tuning approaches, semiconductor and MEMS tuning techniques offer unique advantages and differences. While traditional methods often rely on physical movements or adjustments of mechanical structures, semiconductor and MEMS tuning leverage the properties of semiconductor devices and micrometer-scale mechanical structures. This transition brings forth numerous benefits, including faster response times, precise control, higher integration, and smaller form factors. This section will explore semiconductor and MEMS-based tuning technologies, aiming to gain a deeper understanding of their principles, applications, and advantages.

4.2.1 Semiconductor tuning methods

Semiconductor tuning methods are essential techniques utilized to adjust the electrical properties of semiconductor devices in RF and microwave circuits.

These methods enable precise control over various aspects of circuit performance, including frequency tuning, impedance matching, and power control. Some common semiconductor tuning methods include varactor diodes, field-effect transistors (FETs), PIN diodes, etc. These semiconductor tuning methods are indispensable for the design of RF and microwave circuits, offering adjustable performance parameters that empower engineers to optimize circuit performance for specific applications and operating conditions.

Figure 4.7 gives an example of varactor based tunable RF filter [7]. The filter is based on a dual-behavior resonators topology, employing three pairs of dual-behavior resonators to achieve a third-order bandpass response. At the end of each stub, two varactors are implemented to tune the filter performance. Varactors are semiconductor devices with variable capacitance controlled by varying the applied voltage. In this design, SMV1405-079LF varactors are selected, offering tuning capacitance ranging from 0.63 pF to 2.67 pF over a biasing voltage range of 0 V to 30 V. The top and bottom stubs are controlled by biasing voltages V_1 and V_2, respectively. By adjusting V_1 and V_2, this filter provides high tuning ranges for both center frequency and bandwidth, spanning from 1.59 GHz to 2.7 GHz and from 57 MHz to 481 MHz, respectively, at the middle frequency of 2.24 GHz. It achieves a maximum insertion loss of 5.9 dB.

Varactor-based tuning techniques offer wide tuning ranges and fast tuning speeds due to their ability to change capacitance with applied voltage. They

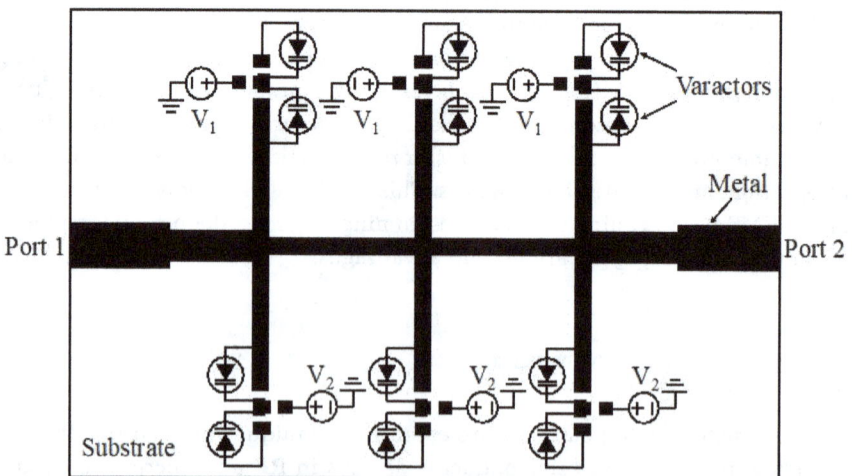

Figure 4.7: A varactor based tunable RF filter.

are also compact and have relatively low power consumption. However, they suffer from nonlinear tuning behavior, limited quality factor, temperature sensitivity, and voltage sensitivity. Another alternative reconfigurable technique is based on field-effect transistors (FETs). FETs operate by controlling the flow of current between their source and drain terminals using an electric field applied to the gate terminal. FETs come in various types, including metal-oxide-semiconductor FETs (MOSFETs) and high-electron-mobility transistors (HEMTs), each offering different performance characteristics suited for specific applications. Compared to varactors, FETs can provide linear tuning behavior, higher quality factors, and better temperature stability. Additionally, FETs offer the design possibility of more complex reconfigurable circuits, enabling functionalities beyond simple frequency tuning.

A compact frequency-reconfigurable antenna based on GaAs FET switch is depicted in Figure 4.8 [8]. The antenna is fabricated on an RO4350B substrate with a relative permittivity of 3.48. It consists of a 50 Ohm microstrip feed line, two strips, a rectangular ring, a switch land pattern, and one of the driving lines located on the top side. On the bottom side, there are partial ground planes and two switch driving lines with pads. The driving lines are connected to the switch land pattern on the top side of vias. The switch control signal can be fed

Figure 4.8: Frequency reconfigurable antenna using GaAs FET switch: (a) top view, (b) bottom view.

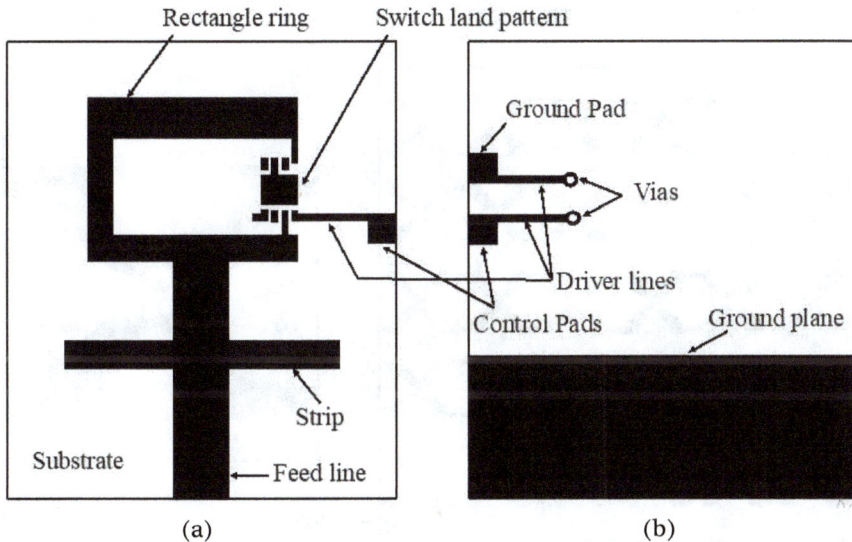

to the driving lines to activate or deactivate the GaAs FET switch. Positioned at the right-hand side of the rectangular ring, the switch alters the radiator shape. When the switch is on, the complete rectangular ring functions as a monopole patch antenna, achieving a narrowband of 2.3–3 GHz. Conversely, when the switch is off, the rectangular ring can be divided into two parts: an asymmetric U-shaped patch and a short strip. This configuration contributes to dual bands of 1.65–2 GHz and 3.4–3.97 GHz, respectively.

Another widely used semiconductor device utilized in RF and microwave circuits is the PIN diode, which consists of three layers—P-type intrinsic and N-type. Its unique structure provides variable capacitance characteristics, enabling it to function as a switch or variable resistor. In comparison to varactors and FETs, PIN diodes offer several advantages such as fast switching speeds, high power handling capability, high quality factor, wideband response, and good temperature stability.

As illustrated in Figure 4.9, a reconfigurable frequency-selective surface (FSS) is enabled by PIN diodes [9]. The unit cell of this FSS consists of a four-armed star geometry and a PIN diode. The connection of the top two arms and the bottom two arms is controlled by the OFF or ON states of the PIN diode. This four-armed star geometry is polarization dependent. Under ideal conditions, if the electric field is horizontally polarized, the OFF and ON states do not affect the frequency response. However, if the electric field is vertically polarized,

Figure 4.9: Reconfigurable frequency selective surface based on a PIN diode.

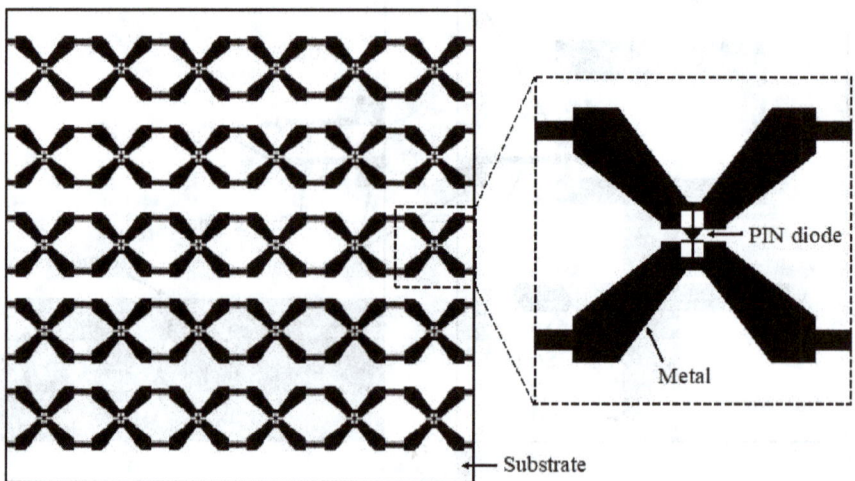

the resonant frequency in the OFF state will be approximately twice that in the ON state. Since the unit cells are connected in rows, the PIN diodes inside each unit cell can be biased by applying voltage on the left or right side. With the OFF and ON states of the PIN diodes, operating frequencies of 6.72 GHz and 4.12 GHz are achieved. Additionally, the design incorporates the PIN diode threshold region, leading to a third situation where the FSS becomes practically transparent, which is an interesting feature.

In summary, semiconductor tuning methods offer versatile techniques for adjusting the performance of RF and microwave circuits. Varactor-based tuning provides wide tuning ranges and fast tuning speeds, suitable for applications requiring rapid frequency adjustments. FET-based reconfigurable techniques offer linear tuning behavior, higher quality factors, and better temperature stability, enabling more precise control over circuit parameters. PIN diodes provide fast switching speeds, high-power handling capability, and wideband response, making them ideal for applications requiring robust performance across a broad frequency range. Each semiconductor tuning method has its advantages and is suitable for specific applications, offering engineers flexibility in designing RF and microwave circuits tailored to their requirements.

4.2.2 MEMS switches

Micro-electromechanical systems (MEMS) refer to miniaturized devices that integrate electrical and mechanical components on a microscopic scale. These systems typically consist of microscale sensors, actuators, and electronics, fabricated using semiconductor manufacturing techniques. MEMS devices utilize the principles of microfabrication to create structures and mechanisms with dimensions ranging from micrometers to millimeters. MEMS technology enables the development of highly compact, low-power, and cost-effective devices with a wide range of applications across various industries.

RF MEMS switches are devices that use mechanical movement to achieve a short circuit or an open circuit [10]. The forces required for the mechanical movement can be obtained using electrostatic, magnetostatic, piezoelectric, or thermal activations. To date, electrostatic-type switches have been demonstrated at frequencies ranging from 0.1 GHz to 100 GHz with high reliability (100 million to 100 billion cycles) and wafer-scale manufacturing techniques. The physical structure of electrostatic-type MEMS switching devices is shown in Figure 4.10. A thin metal membrane is suspended a short distance above a conductor. When a DC potential is applied between the two conductors, charges are induced on the metal, attracting the two electrodes. Above a certain threshold voltage, the force of attraction is sufficient to overcome the mechanical stresses in the membrane material, causing the membrane to snap down to close contact with the bottom conductor.

Figure 4.10: Functional diagrams of two common RF MEMS switch structures: (a) air bridge switch, (b) cantilever switch.

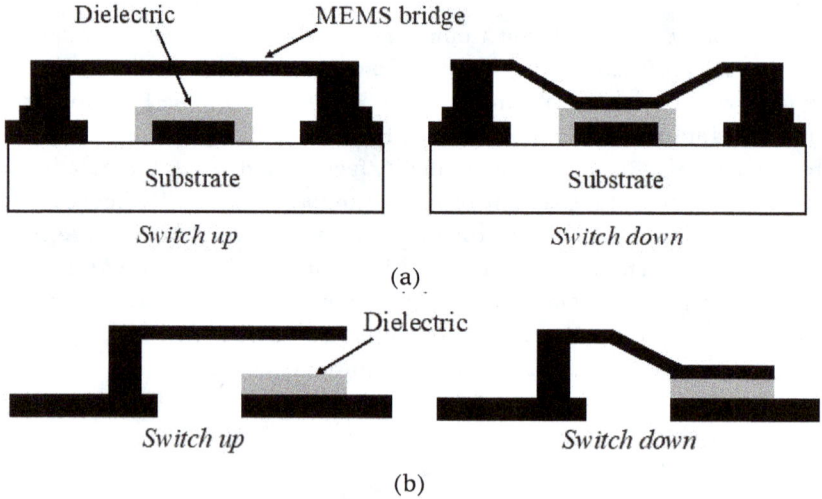

Reconfigurable techniques based on MEMS switches offer versatile solutions for adjusting the performance of RF and microwave circuits. Leveraging MEMS technology, these techniques dynamically alter the physical structure or properties of circuit components. MEMS switches can adjust the resonant frequency of RF components like resonators, filters, and antennas, enabling precise frequency tuning in wireless communication systems. They can also control bandwidth by altering the dimensions or configuration of MEMS-based components such as transmission lines or filters to meet specific system requirements. MEMS-based phase shifters and phased array antennas facilitate electronic beam steering, crucial for radar and satellite communication systems. Moreover, MEMS switches and variable capacitors provide dynamic switching between different circuit configurations or parameter adjustments for adaptive RF circuit designs. Tunable matching networks optimize power transfer with MEMS-based impedance matching, while MEMS-based switches and attenuators offer high-power handling capability and isolation for efficient signal routing and power control in RF systems.

A Ka-band phase shifter using MEMS shunt switches is shown in Figure 4.11 [11]. Since MEMS capacitive membrane switches have low losses and low parasitic effects at frequencies through 40 GHz, the Ka-band phase shifter using these switches can achieve very low RF losses. The operation of the phase shifter is straightforward: if a shunt switch is activated in the reference path, it results in an open circuit at the T-junction, forcing the energy to flow through the delay network. Due to the quarter-wavelength lines used, the single-pole double-throw (SPDT) bandwidth is around ±10%. The microstrip line is kept at DC ground, and 100 kΩ resistors are used to bias the MEMS switches. The average insertion loss is 2.2 dB, with an associated bandwidth of 32 GHz to 36 GHz ($S_{11} < -10$ dB). For different working states, the phase can be changed from 0° to 337.5° with 22.5° steps and the phase error is within 13° for all phase states.

Figure 4.11: A Ka-band phase shifter controlled by MEMS switches.

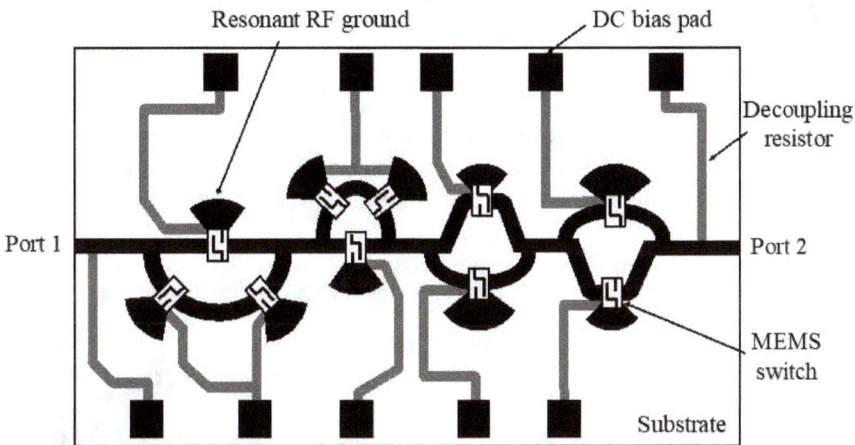

RF MEMS can also be applied to design filters with reconfigurable performance [12]. As shown in Figure 4.12, the filter layout consists of a single resonator with a coupled-feed scheme. Varactors are loaded to provide variable capacitance while an RF MEMS switch is employed to control the connection between the two feed lines. Two capacitors are located at the end of the top resonator to fine-tune the matching, and resistors are introduced for biasing the varactors. The filter is initially designed with a passband centered at 1.12 GHz and a 3-dB fractional bandwidth of 32%. When the MEMS switch is turned ON, a direct path from port 1 to port 2 is created, leading to a band stop filtering response. By contrast, when the MEMS switch is turned OFF, the two feed lines are disconnected, allowing only the signal at the resonant frequency to be coupled and propagated, resulting in a bandpass response. Additionally, the filter bandwidth and transmission zeros/poles can be tuned by changing the capacitance of the varactors.

Figure 4.12: An RF MEMS based bandpass to bandstop switchable filter.

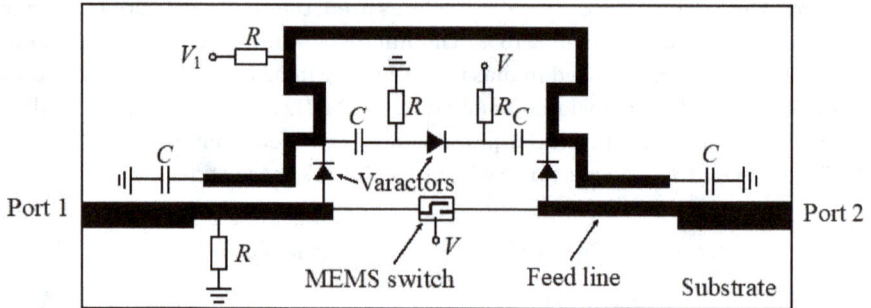

Another example of a reconfigurable antenna using an RF MEMS switch is illustrated in Figure 4.13 [13]. This antenna consists of two layers fabricated on a substrate, with air serving as the dielectric between them. The bottom layer,

Figure 4.13: An RF MEMS-integrated reconfigurable antenna: (a) top view, (b) bottom view.

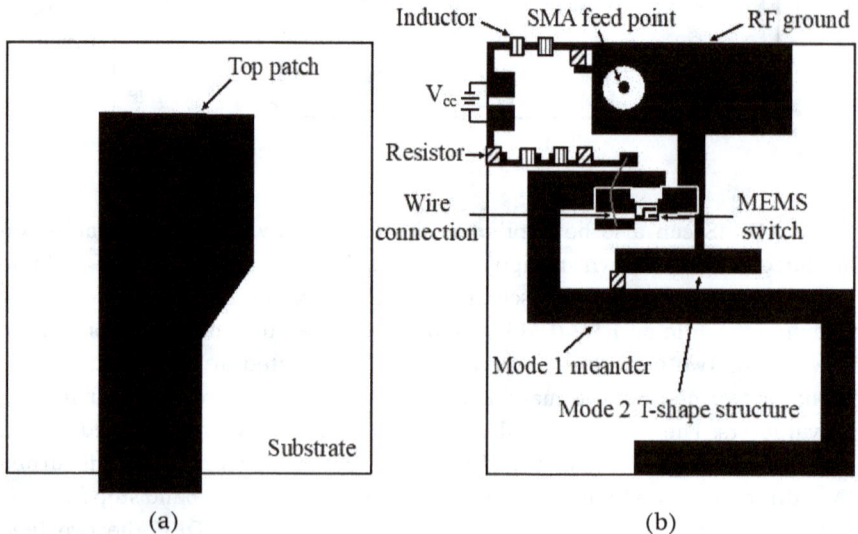

which includes the RF MEMS switch, functions as the reconfigurable metal layer, while the top layer supports the top patch antenna. To incorporate the capacitive feed design, which counteracts the inductive effect of the coaxial feed, the substrate is sandwiched between the bottom conductive plate of the capacitive feed and the top patch metal of the antenna. Inductors and resistors are employed to form the biasing circuit for the RF MEMS switch. The antenna operates in two distinct modes: Mode 1, centered around 718 MHz, when the meander is connected (switch is ON), and Mode 2, centered around 4960 MHz, when the meander is disconnected (switch is OFF).

Overall, MEMS-based reconfigurable techniques provide a flexible and compact solution for dynamic control and optimization of RF and microwave circuits across various applications in wireless communication and radar systems.

4.3 Thin Film Technologies

4.3.1 Ferroelectric and ferromagnetic thin films

Tuning techniques based on thin films, particularly ferroelectric and ferromagnetic thin films, have garnered significant attention in recent years due to their ability to enhance the performance and functionality of various electronic and RF components.

Ferroelectric materials, a subset of piezoelectric and pyroelectric materials, constitute a class of dipolar dielectric materials characterized by a spontaneous electrical polarization that can be re-oriented by an electric field. Examples of ferroelectric materials include lead zirconate titanate (PZT), barium titanate ($BaTiO_3$), strontium titanate ($SrTiO_3$), lithium niobate ($LiNbO_3$), potassium dihydrogen phosphate (KH_2PO_4), etc. As depicted in Figure 4.14 (a), ferroelectric materials contain numerous tiny domains or crystals functioning as electric dipoles. A dipole signifies the presence of separate positive and negative charges within the crystal structure. Upon exposure to an electric field, these dipoles align or orient themselves in a consistent direction, thereby imparting the material with distinct positive and negative sides. Figure 4.14 (b) illustrates the hysteresis loop of ferroelectric materials to describe the relationship of polarization (P) and electrical field (E). With the applied external electric field, the polarization increase nonlinearly and eventually saturates at P_s. The polarization does not disappear when the external electric field is removed, and exhibits a remnant polarization P_r. A corrective filed E_c is required to bring the polarization back to zero.

Figure 4.14: (a) The polarization of ferroelectric materials can be re-oriented by an external electric field. (b) Hysteresis loop of ferroelectric materials.

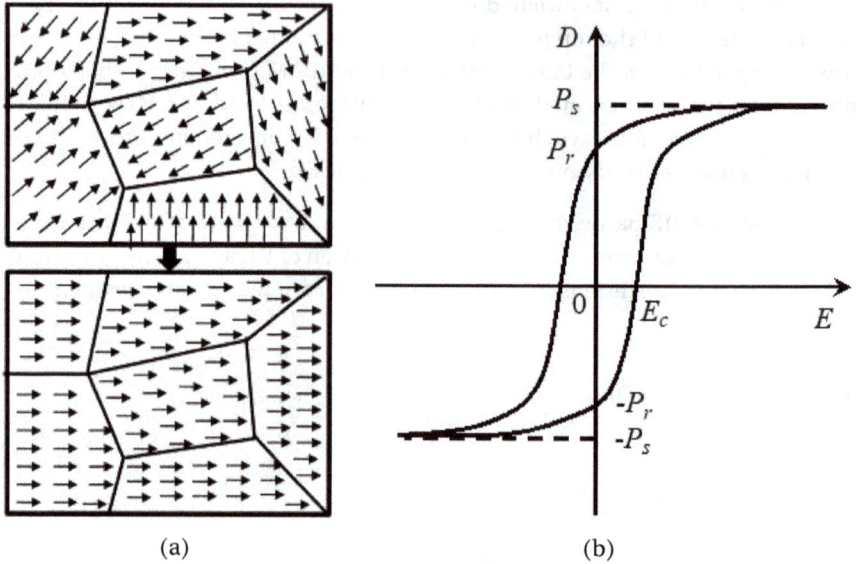

(a)　　　　　　　　　　(b)

The permittivity of ferroelectric materials is the measure of their ability to store electrical energy in an electric field and is of primary consideration in RF designs. Reconfigurable RF technologies leverage modifications in material permittivity to achieve the desired functionalities. Permittivity is defined as the ratio of the electric displacement D to the electric field E ($\epsilon = D/E$). The nonlinear relationship between polarization and the applied electric field is the key factor that allows the permittivity of the material to be tuned. In addition, the ease of applying electric fields within RF circuits enhances the feasibility and promise of utilizing ferroelectric material-based tuning techniques. This accessibility to the applied field not only simplifies the implementation process but also contributes to the scalability and adaptability of RF devices.

While ferroelectric materials offer voltage dependent tunable permittivity, it's also noteworthy to apply ferromagnetic materials. Ferromagnetic materials, similar to ferroelectric ones, exhibit unique magnetic properties that can be manipulated for RF and microwave applications. As depicted in Figure 4.15 (a), when the ferromagnetic material is exposed to a magnetic field, magnetic domains can be aligned in the same direction as the bias magnetic field and will maintain magnetic properties even in the absence of the applied magnetic field. Iron (Fe), nickel (Ni), cobalt (Co) and their alloys are some examples of ferromagnetic materials that are greatly desired in the design of RF components. When a ferromagnetic material is subjected to a magnetic field, the hysteresis loop, depicting the relationship between magnetic field strength (H) and magnetic flux density (B), is illustrated in Figure 4.15 (b). As the external magnetic field increases positively, the magnetic flux density nonlinearly rises until reaching saturation at B_s. Subsequently, upon removing the external magnetic field, the magnetic induction reduces to B_r. Conversely, decreasing the applied magnetic field in the negative direction results in a reduction of magnetic induction to zero, where the necessary magnetic field is termed coercivity (H_c). If the external magnetic field continues to decrease and eventually reversed, the curves symmetrically align to the origin point.

Figure 4.15: (a) The magnetic dipoles within ferromagnetic materials can align in the same direction by an external magnetic field. (b) Hysteresis loop of ferroelectric materials.

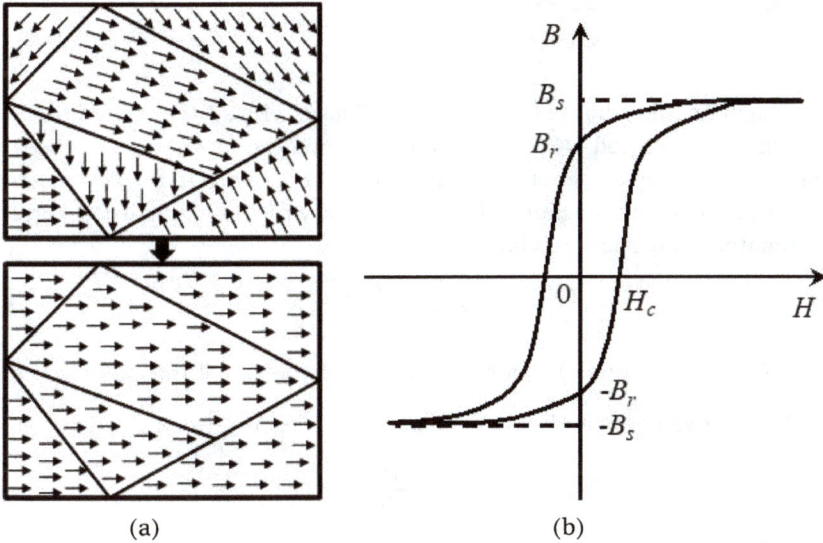

(a)　　　　　　　　(b)

Similarly, the permeability of ferromagnetic materials, defined as the ratio of the magnetic flux density B to the magnetic field strength H ($\mu = B/H$), can be tuned by adjusting the external magnetic field. However, the required external magnetic field is often bulky, consumes relatively high power, and is difficult to integrate with RF systems. To reduce the need for a bulky external magnetic bias field, a strategy has been developed utilizing the static magnetic field generated from the applied DC current to electrically tune the permeability of

Figure 4.16: Electrical tuning method for ferromagnetic thin films.

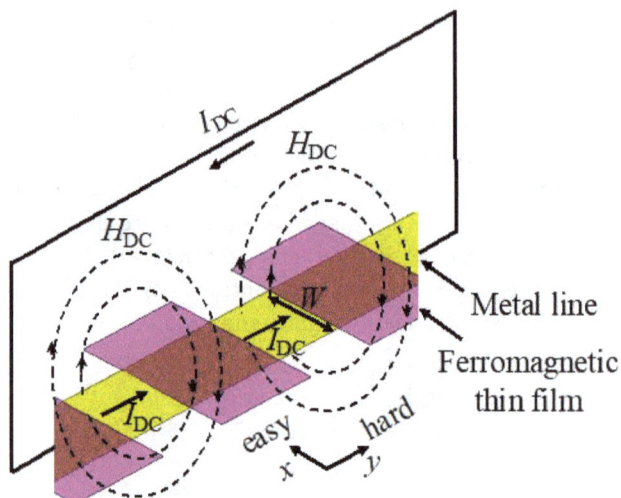

ferromagnetic thin films [14]. As shown in Figure 4.16, a DC current is applied through the patterned DC bias lines, which tune the magnetization distribution inside the ferromagnetic thin films with the associated static magnetic field. The maximum magnetic field associated with the applied DC currents is estimated using Ampere's law:

$$H_{DC} = \frac{I_{DC}}{2W} \tag{4.1}$$

where I_{DC} is the applied DC current and W is the width of the metal bias line.

The relative permeability of ferromagnetic films is given as:

$$\mu_r = \frac{4\pi M_s}{H_k + H_{DC}} + 1 \tag{4.2}$$

where M_s and H_k represent the saturation magnetization and internal induced magnetic field of the ferromagnetic film, respectively. The static magnetic field generated by the DC current tilts the magnetization direction in the film away from its easy axis toward the hard axis, which in turn decreases the material's magnetic moments: saturation magnetization and anisotropy field. Consequently, the equivalent permeability of the ferromagnetic thin film is adjusted accordingly.

There are two additional limitations when applying ferromagnetic thin films to reconfigurable RF and microwave designs: ferromagnetic resonance (FMR) frequency and magnetic loss. Ferromagnetic resonance is the coupling between an electromagnetic wave and the magnetization of the medium through which it passes, resulting in peak loss at the FMR frequency. Ferromagnetic films have been explored for developing RF components with inductive tuning capability; however, the FMR frequency of unpatterned ferromagnetic films typically falls into the sub-GHz range which limits their applications in RF and microwave frequency bands. The resonant frequency of a ferromagnetic film with a parallel applied external field B is given by the Kittel formula [15]:

$$f_{FMR} = \frac{\gamma}{2\pi}\sqrt{(H_{Bias} + H_{Ani} + (N_y - N_z)4\pi M_s)(H_{Bias} + H_{Ani} + (N_x - N_z)4\pi M_s)}$$

(4.3)

where N_x, N_y, and N_z are demagnetization coefficients for the ferromagnetic materials in different directions, M_s is the saturation magnetization, γ is the gyromagnetic ratio, H_{Bias} is the applied magnetic field, and H_{Ani} is the self-biased shape anisotropy field.

Patterning ferromagnetic thin film with a high aspect ratio is an efficient and flexible strategy to increase its FMR for high frequency applications with reduced losses. Properly patterned ferromagnetic thin films have built-in high

Figure 4.17: Measured FMR frequencies of patterned permalloy thin films with various length to width aspect ratios.

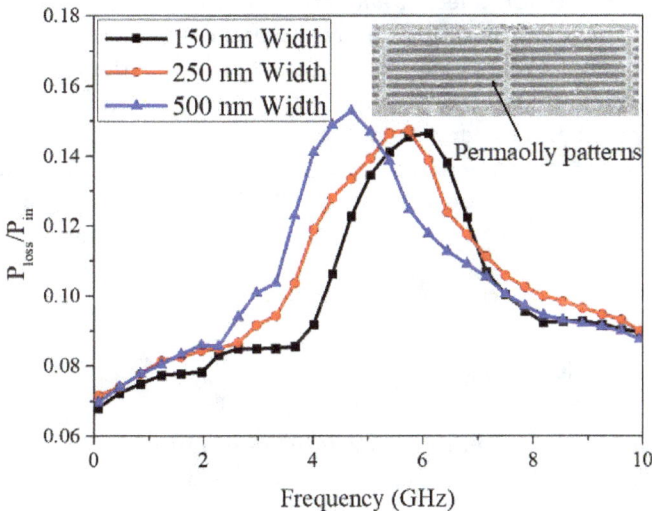

shape anisotropy fields providing a self-biasing field, thus increasing the FMR frequency according to Kittel's equation. Figure 4.17 illustrates the power loss of transmission lines containing permalloy ($Ni_{80}Fe_{20}$) thin film patterns with various aspect ratios [16]. The length of the permalloy patterns and the gap between adjacent patterns are consistently set at 10 μm and 100 nm, respectively. The findings showcase that the attainment of a FMR frequency exceeding 6 GHz through the patterning of permalloy films into nanoscale stripes with substantial aspect ratios, achieved by altering the width from 500 nm to 150 nm. Patterned ferromagnetic thin films, such as ferrite, and CoNbZr, exhibiting higher saturation magnetic fields, hold potential for applications surpassing 20 GHz.

Besides the limitation of FMR frequency, another challenge for the application of ferromagnetic materials is their magnetic losses. Energy dissipation in these materials occurs through two primary mechanisms: hysteresis loss and eddy current loss. Hysteresis loss, as illustrated in Figure 4.15 (b), is the energy dissipated as heat due to the hysteresis effect. For effective use in RF components, materials with a narrow hysteresis loop are preferred. Eddy current loss, on the other hand, arises from the energy dissipated by free electrical charges within a conducting material when exposed to a varying magnetic field. Minimizing eddy current loss is crucial for RF applications. Eddy currents induced on the thin film surface can cause resistive losses and reduce effective permeability. To mitigate eddy current losses, the thickness of the thin film is kept below the skin depth, and the thin film patterns are oriented perpendicular to the magnetic direction, thereby reducing agglomerated particles. [17].

In summary, both ferroelectric and ferromagnetic thin films provide unique tuning advantages, offering wide tuning ranges, fast response times, and compatibility with modern fabrication processes. These properties make them valuable components in the development of advanced RF and microwave devices, contributing to improved performance and adaptability in communication systems, sensors, and other electronic applications. In addition, by leveraging both ferroelectric and ferromagnetic materials, RF engineers can explore comprehensive tuning approaches, further expanding the capabilities and potential applications of RF devices.

4.3.2 Thin film enabled tunable passives

Following the discussion on the characteristics of ferroelectric and ferromagnetic thin films, attention now shifts to their application in the development of tunable passive devices. By exploiting the unique properties of these thin

Figure 4.18: Tunable octagon spiral inductor with ferromagnetic thin films.

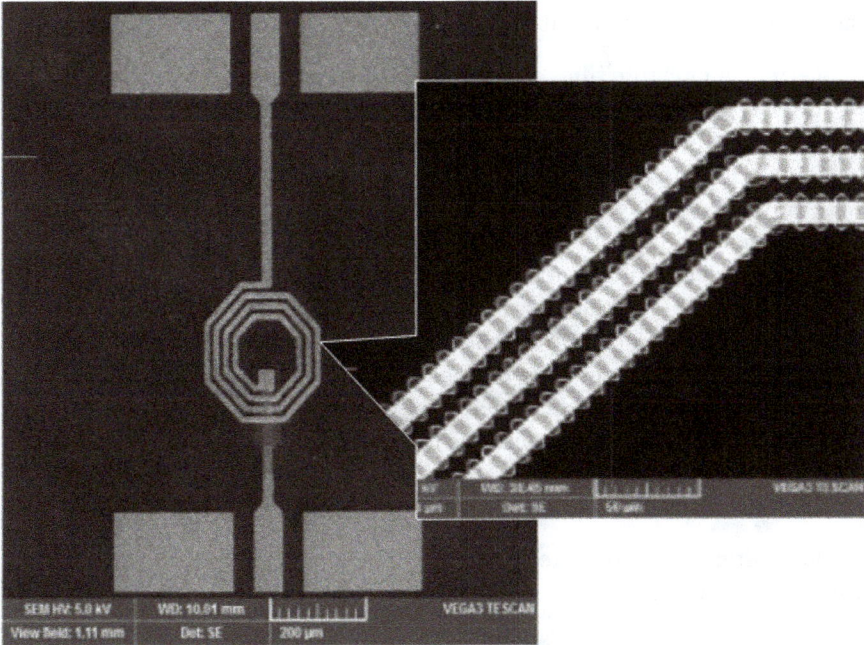

films, it is possible to enhance the performance and functionality of various RF and microwave components. Figure 4.18 shows a tunable octagon spiral inductor enabled with permalloy thin film patterns [18]. Initially, a 1 mm thick gold layer is deposited on the silicon wafer using the photolithography process. Parallel-oriented permalloy is then deposited on top of the metal wires in an atmosphere of 2.1 mT argon gas at room temperature using a DC magnetron gun, introducing high permeability along its hard axis. A thin layer of 5–10 nm chromium is used as an adhesion layer between the gold and the permalloy. Tunable inductors with 100 nm and 200 nm permalloy thin films were fabricated to compare inductance enhancement and tunability. With the application of 100 nm permalloy thin film, the measured inductance of the tunable inductor significantly increased from 9.44 nH to 14.54 nH at 2.2 GHz. On the other hand, 200 nm permalloy thin films have more impact, increasing the inductance from 9.44 nH to 14.83 nH. This improvement is attributed to the high permeability along the hard axis of the permalloy thin film. Additionally, the inductance of the permalloy-enhanced inductor can be tuned from 14.54 nH to 14.20 nH for 100 nm permalloy patterns and from 14.83 nH to 14.25 nH for 200 nm permalloy patterns by applying a static magnetic field generated by DC current. The use of thicker or multiple layers of permalloy thin film further increases the inductance density and tunability.

Ferromagnetic thin films can also be utilized in the design of phase shifters to achieve tunable and enhanced performance. An electrically tunable phase shifter based on a slow-wave coplanar waveguide (CPW) structure is illustrated in Figure 4.19, with an inset showing a close-up view of the permalloy thin films. A 1 Âm thick gold layer is deposited and patterned as a CPW structure on a high-resistivity silicon substrate. The CPW transmission line is designed in a slow-wave configuration, featuring alternating narrow and wide signal sections. Subsequently, 200 nm thick permalloy thin films are patterned on top of the CPW structure using the DC magnetron sputtering method. The introduction of permalloy enables inductive tuning by altering the inductance of the CPW structure through the magnetic field generated by the applied DC current. As a result of this inductance change, the measured working frequency for the 90° phase shift can be continuously tuned from 1.8 GHz to 2 GHz under different applied DC currents.

In addition to utilizing ferroelectric or ferromagnetic thin films individually, it is also possible to integrate both types of thin films for simultaneous application in the design of RF devices. This hybrid approach can leverage the unique properties of each material, potentially leading to enhanced performance, greater tunability, and improved functionality in RF components. Figure 4.20 depicts a tunable lumped elements phase shifter with both ferroelectric and ferromagnetic thin films [20]. The phase shifter is constructed using a 3D solenoid with a specially designed magnetic core and a metal–insulator–metal (MIM) capacitor. The magnetic core is created by two layers of patterned permalloy thin films and a layer of silicon oxide serving as an insulator. The high, tunable permeability of the permalloy patterns significantly increases the inductance of the phase shifter. Additionally, MIM capacitors are employed with lead zirconate titanate (PZT) as the insulator to alter the capacitance of the phase shifter. Permalloy and PZT can be electrically tuned by applying DC current and DC voltage, respectively. Integrating both ferromagnetic and ferroelectric materials provides dual tunability, further enhancing performance and offering a wider tuning range for the phase shifter. The measured results show that the phase shift of the phase shifter is 59.2° at 2 GHz without any DC bias. When a 150 mA DC current is applied, the phase shift changes to 53.8°, representing an inductive tunability of 9.1%. When a 6 V DC voltage is applied, the phase shift adjusts to 48.8°, providing a capacitive tunability of 17.6%. When both DC current and voltage are applied simultaneously, the phase shift further adjusts to 43.8°, achieving a combined dual tunability of 26.9%. These results demonstrate the effectiveness of combining ferromagnetic and ferroelectric

Figure 4.19: A tunable phase shifter enabled with permalloy thin film patterns.

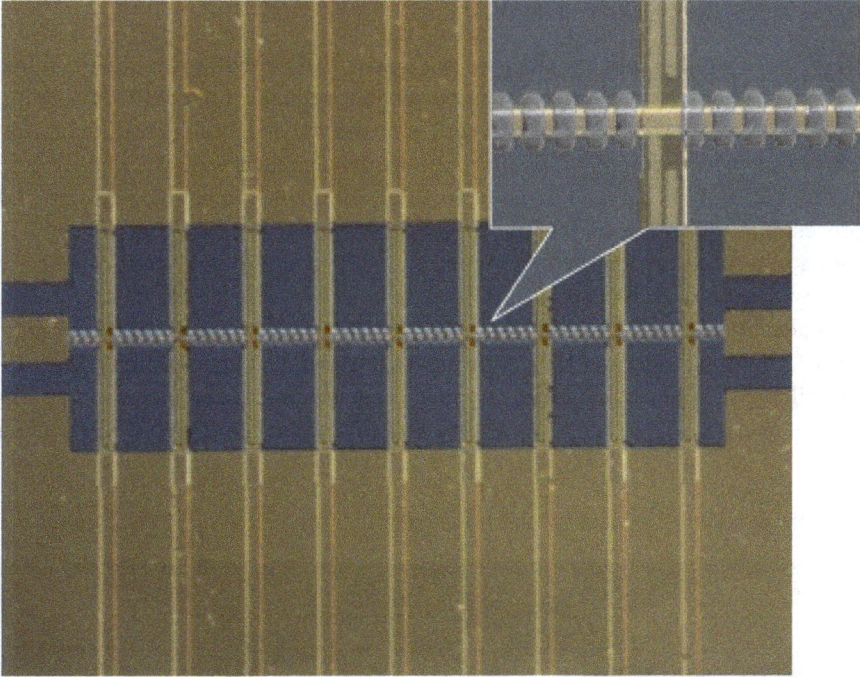

materials to achieve significant tunability in phase shifters, paving the way for more advanced and adaptable RF devices.

Beyond designing specific RF devices with ferroelectric and ferromagnetic thin films, their high permittivity and high permeability characteristics can also be utilized to achieve specific functionalities. For instance, these properties can be employed to reduce far-end crosstalk between coupled lines in integrated circuits. Crosstalk refers to the coupling of energy from one signal to another, occurring due to the interaction of electric and magnetic fields between transmission lines. Far-end crosstalk (FEXT) is defined as the difference between the capacitive coupling and the inductive coupling between two adjacent signal lines. The FEXT between two closed transmission lines is calculated as:

$$FEXT = \frac{V_{in}l}{RT} \times \frac{1}{2v} \times \left(\frac{C_m}{C_s} - \frac{L_m}{L_s} \right) \tag{4.4}$$

where V_{in} is the input voltage, l is the line length, RT is the signal rise time, v is the speed of the signal on the line, C_m is the mutual capacitance per length, C_s

is the self-capacitance per length, L_m is the mutual inductance per length, and L_s is the self-inductance per length. It is noted that FEXT increases with both signal frequency and trace length.

For two coupled microstrip lines, the capacitive coupling is usually smaller than the inductive coupling, resulting in a negative magnitude of FEXT. Adding short trapezoidal-shaped tabs to the edges of the traces can increase the mutual capacitance, reducing the difference between the capacitive and inductive coupling [21]. However, FEXT cannot be fully eliminated due to geometric and process limitations in further increasing the mutual capacitance. Figure 4.21 shows two solutions for further reducing FEXT by adding either ferroelectric or ferromagnetic thin films [22]. Adding a layer of ferroelectric thin film with high permittivity between the two traces can further increase the mutual capacitance, thereby reducing FEXT. Conversely, patterning a layer of ferromagnetic thin film with high permeability on top of the metal traces can increase the self-inductance, which also reduces the difference between capacitive and inductive

Figure 4.20: A tunable phase shifter enabled by both permalloy and PZT thin films.

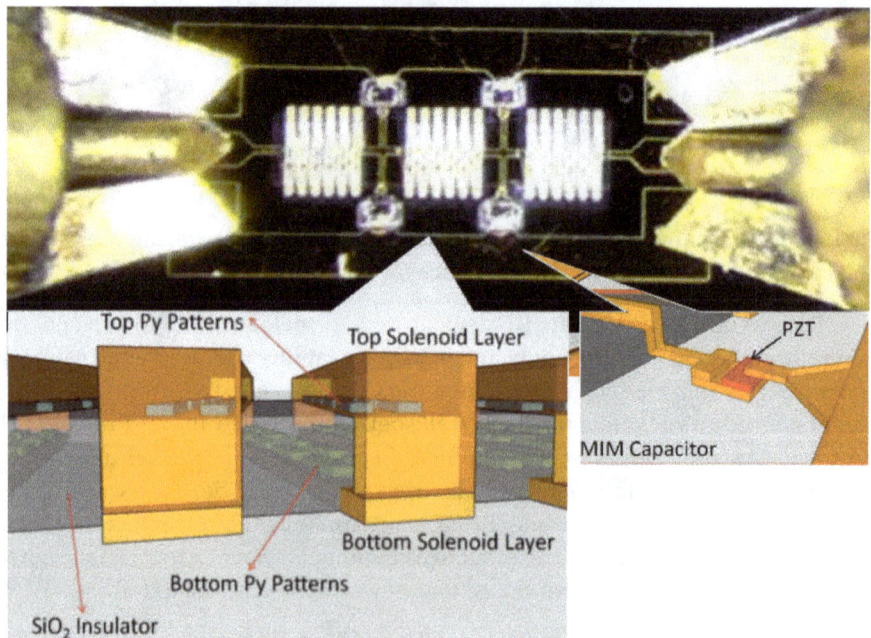

Figure 4.21: Far end crosstalk reduction for coupled lines by applying ferroelectric and ferro-magnetic thin films.

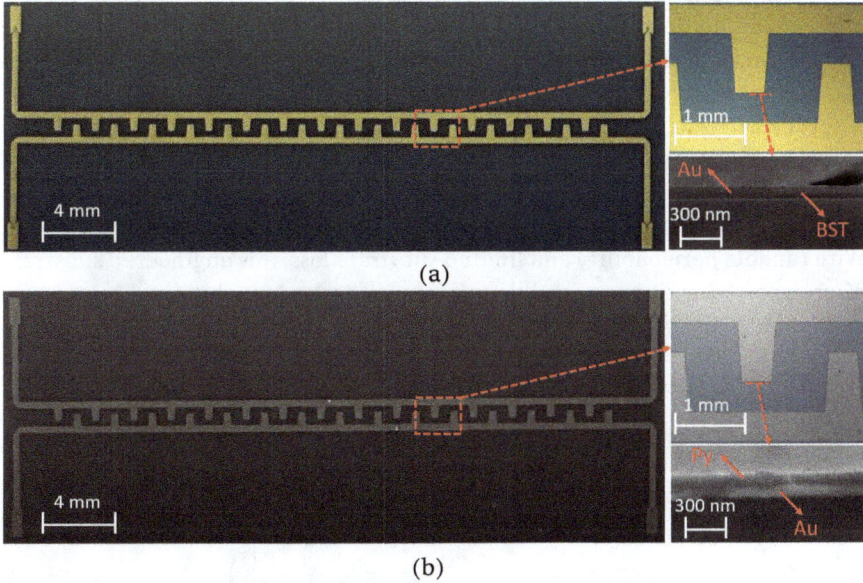

(a)

(b)

coupling, leading to improved FEXT. Additionally, applying a layer of ferromag-netic thin film can help reduce impedance variation along the tabbed traces, thereby minimizing mismatch loss.

4.3.3 Thin film enabled engineered substrate

In modern wireless communication systems, most miniaturized RF passives are implemented on dielectric substrates with high permittivity. Typically, high permittivity dielectrics cause significant capacitive effects between the metal traces and the reference ground. To mitigate these coupling effects and main-tain the desired RF impedance, high inductance from narrow traces is required to balance the high capacitance. However, this approach can lead to increased loss and coupling noise between traces. Magneto-dielectrics, which exhibit both high permittivity and high permeability, offer a promising solution by reducing capacitive effects while maintaining desirable RF performance [23].

Magneto-dielectric substrates are engineered to enhance specific electro-magnetic properties, making them suitable for advanced RF and microwave applications. Ferromagnetic materials are commonly utilized in magneto-dielectric substrates to achieve high permeability. However, they often suffer from high magnetic loss and low FMR frequency. As previously discussed, properly patterning ferromagnetic thin films with a high aspect ratio can increase the FMR frequency and reduce magnetic loss, making them suitable for forming magneto-dielectric substrates for RF applications. Figure 4.22 illustrates a specifically engineered substrate with patterned ferromagnetic thin films [24]. The engineered substrate consists of a standard RF substrate, patterned ferromagnetic thin films, and metal bias lines. By applying DC current to the metal bias lines, the permeability of the engineered substrate can be tuned. With tunable permeability and limited substrate loss, this engineered substrate offers a cost-effective and flexible solution for designing arbitrary miniaturized and tunable RF components.

Figure 4.22: Engineered substrate enabled by ferromagnetic thin films.

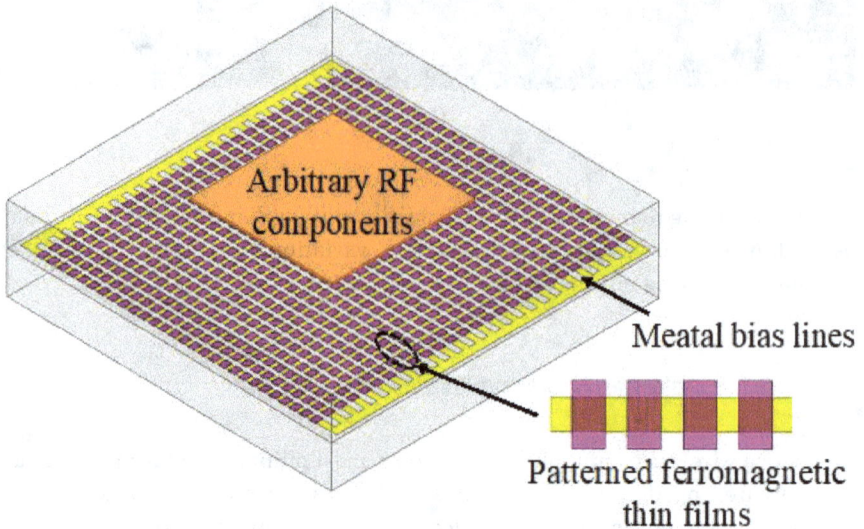

Higher permeability of the engineered substrate can be achieved by implementing more ferromagnetic thin film layers and the performance of the engineered substrate is characterized in [25]. As shown in Figure 4.23, a microstrip line is implemented on the engineered substrate and an equivalent model with effective permeability ($m_{r,eff}$) is established to mimic the performance of the real model. When the extracted inductance of both models is the same, the effective permeability of the engineered substrate can be derived. After exploring the impacts of multiple factors, such as film thickness, number of film layers, film vertical position, planar filling density, and pattern dimensions, an accurate model is developed to quickly optimize and determine the configuration of the engineered substrate for the desired permeability.

Figure 4.23: Equivalent model of the engineered substrate with ferromagnetic thin films.

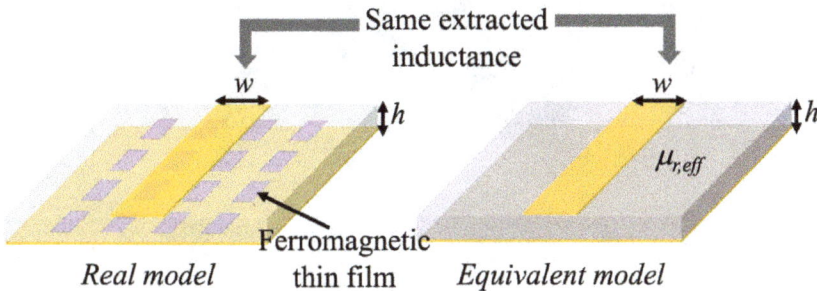

With the developed model, the engineered substrate can be easily applied to RF designs. A good example is a miniaturized and tunable frequency selective surface with multiple layers of the ferromagnetic thin films [26]. The engineered substrate used in this design is based on the measured results of a sample substrate with one layer of permalloy thin film in [27]. The permalloy thin film is patterned as an array of 15 mm × 20 μm with 10 μm gaps among them. The measured permeability of the engineered substrate is 1.140 and can be tuned to 1.102 by applying a DC current of up to 500 mA. Using the aforementioned engineered substrate model, a higher permeability of 2.398 can be achieved by increasing the number of permalloy thin film layers to 10. Based on this, an FSS consisting of periodic square rings is implemented on the engineered substrate, as depicted in Figure 4.24 (a). Due to the increased permeability of the engineered substrate, this FSS achieves 16% size reduction compared to the FSS on a normal substrate. In addition, as shown in Figure 4.24 (b), the operating frequency of the miniaturized FSS is tuned from 2.45 GHz to 2.672 GHz by applying DC current from 0 mA to 500 mA.

The increased permeability of the engineered substrate also brings great benefits to antenna designs, including further miniaturization, wider bandwidth, and higher efficiency. Figure 4.25 shows a patch antenna on the engineered substrate with 10 layers of permalloy thin films [28]. The antenna is first designed on a standard substrate with an operating frequency of 2.45 GHz to serve as a reference. By contrast, increasing the effective permeability of the engineered substrate to 2.398, leads to a smaller miniaturization factor

Figure 4.24: (a) A square ring based FSS on the engineered substrate. (b) Frequency response of the miniaturized FSS with different biasing currents.

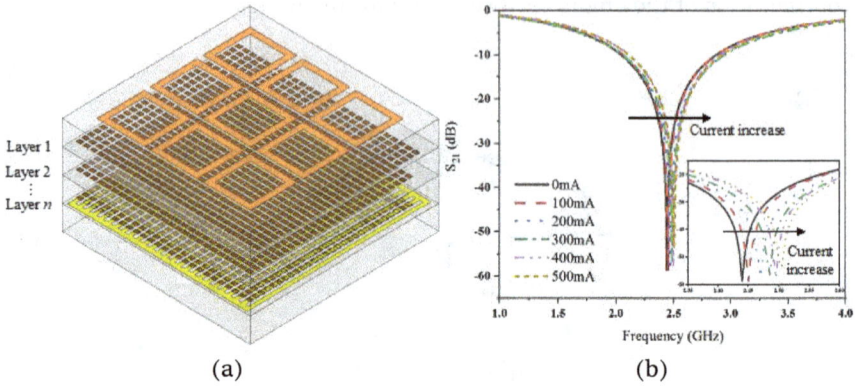

(a)   (b)

Figure 4.25: (a) A patch antenna on the engineered substrate. (b) Performance comparison of the antenna on a normal dielectric substrate and the engineered substrate.

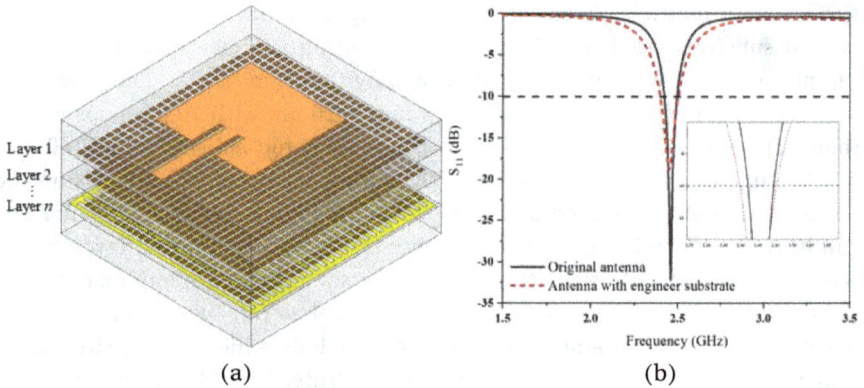

(a)   (b)

$(\sqrt{\varepsilon_r \mu_r})$. Therefore, a 47.6% size reduction can be achieved by designing the antenna on the engineered substrate. Conversely, if the miniaturization factor is maintained as in the original design, the engineered substrate will require a lower permittivity, resulting in less capacitive coupling between the antenna and the ground. This reduction in capacitive coupling significantly benefits the antenna bandwidth.

For a patch antenna, the zero-order bandwidth is approximated by [29]:

$$BW = \frac{96\sqrt{\frac{\mu_r}{\varepsilon_r}} \times \frac{h}{\lambda_0}}{\sqrt{2}\left(4 + 17\sqrt{\mu_r \varepsilon_r}\right)} \tag{4.5}$$

where h is the thickness of the substrate. It is found that the antenna bandwidth can be enhanced by increasing the ratio of μ_r/ϵ_r for a given miniaturization factor. Compared to the original antenna on a conventional dielectric substrate, the same miniaturization factor can be achieved by the engineered substrate with lower permittivity, leading to a wider bandwidth, as shown in Figure 4.25 (b). Furthermore, the increased permeability and decreased permittivity make the antenna impedance close to the air impedance, leading to a high radiation efficiency.

If the engineered substrate is applied to an ultra-wideband (UWB) antenna, the advantage of increased bandwidth will be even more pronounced. As shown in Figure 4.26 (a), the UWB antenna in [30] is co-designed with the engineered substrate. When the permeability of the engineered substrate is selected to equal its permittivity, the substrate impedance becomes close to the free-air impedance, resulting in the best energy radiation efficiency. Figure 4.26 (b) gives the return loss comparison of the UWB antenna with and without the engineered substrate. The engineered substrate is designed to have the same permittivity and permeability of 2.1. Compared to the original antenna with a normal dielectric substrate, the -10 dB bandwidth of the antenna is significantly increased by 50.9%. This design example fully demonstrates the significant benefits of the engineered substrate. Additionally, the permeability of the engineered substrate can be tuned by applying different DC biasing currents, providing greater design flexibility for various antenna functions.

Besides the FSS and antenna designs, the engineered substrate with ferromagnetic thin films has also been applied in designing miniaturized and tunable RF components with enhanced performance and tuning flexibility. Figure 4.27 illustrates two additional designs that primarily utilize the tunable permeability of the engineered substrate. With this tunable permeability, the bandpass filter in [31] and the balun filter in [32] can achieve continuously tunable passbands. Additionally, the increased permeability of the engineered substrate aids in achieving significant miniaturization in these designs.

In summary, the direct application of ferroelectric and ferromagnetic thin films has demonstrated significant benefits, including enhanced performance and electrical tunability for specific RF components. Thin films can

Figure 4.26: (a) Ultra-wide band antenna with engineered substrate. (b) Performance comparison.

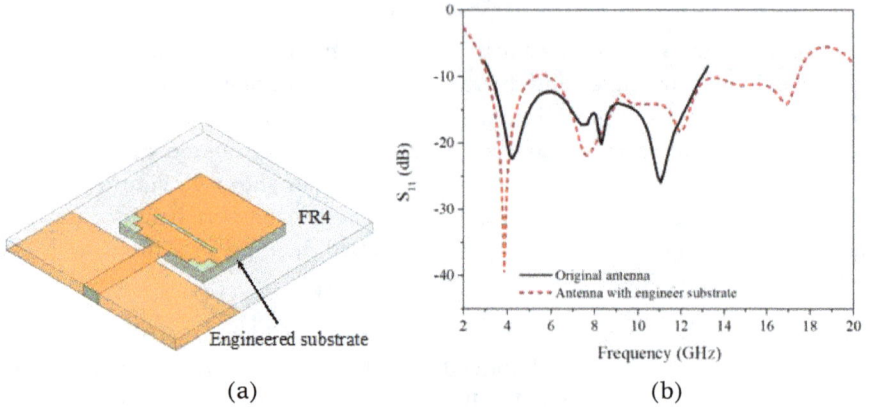

(a)   (b)

Figure 4.27: RF designs based on the engineered substrate: (a) tunable filter, (b) tunable balun.

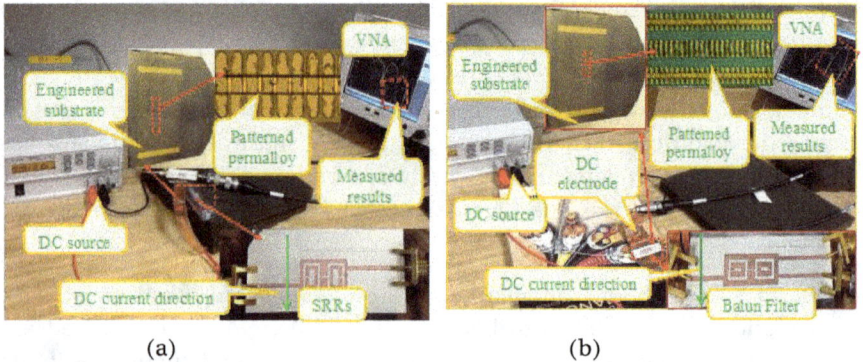

(a)   (b)

also be integrated to form an engineered substrate, resulting in increased and tunable permittivity and permeability. With such tunability, the engineered substrate provides great design flexibility for arbitrary RF designs. Looking forward, the following directions can be pursued for better designs and broader applications: (1) Improving the tuning efficiency and tunability range by investigating the configurations and types of ferroelectric and ferromagnetic thin films. (2) Integrating both ferromagnetic and ferroelectric materials within a design to achieve dual inductive and capacitive tunability.

(3) Developing more integration methodologies for ferroelectric and ferromagnetic thin films to expand their use across a wider range of microwave components.

References

[1]   J. Ge and G. Wang. Switchable Dual-Band Bandpass Filter with High Selectivity. *2021 IEEE MTT-S International Wireless Symposium (IWS)*, 1-3, 2021.

[2]   S. Ohshima et al. Development of High-Speed Mechanical Tuning System for HTS Filters. *IEEE Transactions on Applied Superconductivity*, 19(3), 903-906, 2009.

[3]   F. Mira, J. Mateu and C. Collado. Mechanical Tuning of Substrate Integrated Waveguide Filters. *IEEE Transactions on Microwave Theory and Techniques*, 63(12), 3939-3946, 2015.

[4]   J. Ge and G. Wang. CmWave to MmWave Reconfigurable Antenna for 5G Applications. *2020 IEEE International Symposium on Antennas and Propagation and North American Radio Science Meeting*, 619-620, 2020.

[5]   Weidong Ma, Guangming Wang, Bin-feng Zong, Yaqiang Zhuang and Xiaofei Zhang. Mechanically Reconfigurable Antenna Based on Novel Metasurface for Frequency Tuning-range Improvement. *2016 IEEE International Conference on Microwave and Millimeter Wave Technology (ICMMT)*, 629-631, 2016.

[6]   P. Sanchez-Olivares and J. L. Masa-Campos. Mechanically Reconfigurable Conformal Array Antenna Fed by Radial Waveguide Divider with Tuning Screws. *IEEE Transactions on Antennas and Propagation*, 65(9), 4886-4890, 2017.

[7]   Othman A, Barrak R, Abib G I, et al. A Varactor Based Tunable RF Filter for Multi-standard Wireless Communication Receivers. *AEU-International Journal of Electronics and Communications*, 102, 69-77, 2019.

[8]   X. -l. Yang, J. -c. Lin, G. Chen and F. -l. Kong. Frequency Reconfigurable Antenna for Wireless Communications Using GaAs FET Switch. *IEEE Antennas and Wireless Propagation Letters*, 14, 807-810, 2015.

[9]   Mamedes D F, Gomes Neto A, Costa e Silva J, et al. Design of Reconfigurable Frequency Selective Surfaces Including the PIN Diode Threshold Region. *IET Microwaves, Antennas & Propagation*, 12(9), 1483-1486, 2018.

[10] Wang G. *RF MEMS Switches with Novel Materials and Micromachining Techniques for SOC/SOP RF Front Ends*. 2006.

[11] G. M. Rebeiz, Guan-Leng Tan and J. S. Hayden. RF MEMS Phase Shifters: Design and Applications. *IEEE Microwave Magazine*, 3(2), 72-81, 2002.

[12]   N. Kumar and Y. K. Singh. RF-MEMS-Based Bandpass-to-Bandstop Switchable Single- and Dual-Band Filters with Variable FBW and Reconfigurable Selectivity. *IEEE Transactions on Microwave Theory and Techniques*, 65(10), 3824-3837, 2017.

[13]   A. Zohur, H. Mopidevi, D. Rodrigo, M. Unlu, L. Jofre and B. A. Cetiner. RF MEMS Reconfigurable Two-Band Antenna. *IEEE Antennas and Wireless Propagation Letters*, 12, 72-75, 2013.

[14]   Guoan Wang, Farid Rahman, Tian Xia and Hanqiao Zhang. Patterned Permalloy and Barium Strontium Titanate Thin Film Enabled Tunable Slow Wave Elements for Compact Multi-band RF Applications. *IEEE Transactions on Magnetics*, 49(7), 1-4, 2013.

[15]   C. Kittel. *Introduction to Solid State Physics*, 2004.

[16]   Tengxing Wang. *Integrating Nano-Patterned Ferromagnetic and Ferroelectric Materials for Smart Tunable Microwave Applications*. 2017.

[17]   Rahman, BM Farid, et al. Application of sub-micrometer patterned Permalloy thin film in tunable radio frequency inductors. *Journal of Applied Physics*, 117(17), 17C121, 2015.

[18]   Wang, Tengxing, et al. Integrating Nanopatterned Ferromagnetic and Ferroelectric Thin Films for Electrically Tunable RF Applications. *IEEE Transactions on Microwave Theory and Techniques*, 65(2), 504-512, 2017.

[19]   Rahman, BM Farid, et al. A compact frequency tunable radio frequency phase shifter with patterned Py enabled transmission line. *Journal of Applied Physics*, 117(17), 17D723, 2015.

[20]   T. Wang et al. Novel Electrically Tunable Microwave Solenoid Inductor and Compact Phase Shifter Utilizing Permalloy and PZT Thin Films. *IEEE Transactions on Microwave Theory and Techniques*, 65(10), 3569-3577, 2017.

[21]   S. K. Lee, K. Lee, H. J. Park, and J. Y. Sim. FEXT-eliminated Stub-alternated Microstrip Line for Multi-gigabit/Second Parallel Links. *Electronics Letters*, 44(4), 272–273, 2008.

[22]   Wang G, Ge J. Method and Design of High-Performance Interconnects with Improved Signal Integrity. *U.S. Patent*, 11,744,006. 2023.

[23]   H. Mosallaei and K. Sarabandi. Magneto-dielectrics in Electromagnetics: Concept and Applications. *IEEE Transactions on Antennas and Propagation*, 52(6), 1558-1567, 2004.

[24]   Yujia Peng, Tengxing Wang, Yong Mao Huang, Wei Jiang and G. Wang. Electrically tunable bandpass filtering balun on engineered substrate embedded with patterned Permalloy thin film. *2016 IEEE MTT-S International Microwave Symposium (IMS)*, 1-3, 2016.

[25]   J. Ge and G. Wang. Characterization of Thin Film Enabled Engineered Substrate for RF Applications. *2020 IEEE International Symposium on*

Antennas and Propagation and North American Radio Science Meeting, 769-770, 2020.

[26]   Jinqun Ge, Guoan Wang. Electrically Tunable Miniaturized Band-Stop Frequency Selective Surface on Engineered Substrate with Embedded Permalloy Patterns. *AIP Advances*, 9 (12), 125145, 2019.

[27]   Peng, Yujia, et al. Performance Enhanced Miniaturized and Electrically Tunable Patch Antenna with Patterned Permalloy-based Magneto-dielectric Substrate. *Journal of Applied Physics*, 115(17), 17A505, 2014.

[28]   J. Ge and G. Wang. Thin Film Enabled Engineered Substrate for Miniaturized Antennas with Improved Bandwidth. *2020 IEEE International Symposium on Antennas and Propagation and North American Radio Science Meeting*, 739-740, 2020.

[29]   R. C. Hansen and M. Burke. Antenna with Magneto-dielectrics. *Microwave and Optical Technology Letters*, 26(2), 75–78, 2000.

[30]   Choi, Seok H., et al. A New Ultra-Wideband Antenna for UWB Applications *Microwave and Optical Technology Letters*, 40(5), 399-401, 2004.

[31]   Peng, Yujia, et al. Engineered Smart Substrate with Embedded Patterned Permalloy Thin Film for Radio Frequency Applications. *Journal of Applied Physics*, 117(17), 17B709, 2015.

[32]   Yujia Peng, Tengxing Wang, Yong Mao Huang, Wei Jiang, and G. Wang. Electrically Tunable Bandpass Filtering Balun on Engineered Substrate Embedded with Patterned Permalloy Thin Film. *2016 IEEE MTT-S International Microwave Symposium (IMS)*, 1-3, 2016.

Index

For Product Safety Concerns and Information please contact our EU
representative GPSR@taylorandfrancis.com
Taylor & Francis Verlag GmbH, Kaufingerstraße 24, 80331 München, Germany

www.ingramcontent.com/pod-product-compliance
Lightning Source LLC
Chambersburg PA
CBHW061611220326
41598CB00024BC/3537

* 9 7 8 8 7 7 0 0 4 7 8 8 3 *